UNIVERSITY OF GLASGOW SOCIAL
AND ECONOMIC STUDIES

General Editor: Professor D. J. Robertson

18

LABOUR PROBLEMS
OF TECHNOLOGICAL CHANGE

UNIVERSITY OF GLASGOW SOCIAL
AND ECONOMIC STUDIES

New Series
General Editor: *Professor D. J. Robertson*

1. The Economics of Subsidizing Agriculture. GAVIN MCCRONE
2. The Economics of Physiocracy. RONALD L. MEEK
3. Studies in Profit, Business Saving and Investment in the United Kingdom, 1920–62. Vol. 1. P. E. HART
4. Scotland's Economic Progress. GAVIN MCCRONE
5. Fringe Benefits, Labour Costs and Social Security. Edited by G. L. REID and D. J. ROBERTSON
6. The Scottish Banks. MAXWELL GASKIN
7. Competition and the Law. ALEX HUNTER
8. The Nigerian Banking System. C. V. BROWN
9. Export Instability and Economic Development. ALASDAIR I. MACBEAN
10. The Mines of Tharsis, Roman French and British Enterprise in Spain. S. G. CHECKLAND
11. The Individual in Society: Papers on Adam Smith. A. L. MACFIE
12. The Development of British Industry and Foreign Competition 1875–1914. Ed. DEREK H. ALDCROFT
13. Studies in Profit, Business Saving and Investment in The United Kingdom, 1920–62. Vol. 2. Ed. P. E. HART
14. Building in the Economy of Britain between the Wars. Ed. DEREK H. ALDCROFT and HARRY W. RICHARDSON
15. Regional Policy in Britain. GAVIN MCCRONE
16. Regional and Urban Studies, A Social Science Approach. Ed. SARAH C. ORR and J. B. CULLINGWORTH
17. Planning Local Authority Services for the Elderly. GRETA SUMNER and RANDALL SMITH

LABOUR PROBLEMS
OF TECHNOLOGICAL
CHANGE

BY

L. C. HUNTER
G. L. REID
D. BODDY

Department of Social and Economic Research
University of Glasgow

London
GEORGE ALLEN & UNWIN LTD
RUSKIN HOUSE   MUSEUM STREET

FIRST PUBLISHED IN 1970

© George Allen & Unwin Ltd, 1970

SBN 04 331045 1

PRINTED IN GREAT BRITAIN
in 10 on 11 pt. Times type
BY T. & A. CONSTABLE LTD
EDINBURGH

PREFACE

This book is concerned with the examination and assessment of the impact of changes in technology on companies in three selected industries: printing, steel and chemicals. Its main focus is on the employment and associated labour market effects of technological change; but part of the rationale for the study as a whole has been to relate these effects to the technological environment of each industry. Accordingly a good deal of attention has been paid to the character of the innovations themselves and to their implications for the industries in general terms.

The study may be regarded as something of a 'half-way house', for it neither adopts the macro-economic and often econometric approach to the effects of technological change—which all too often embraces factors which cannot realistically be regarded as part of the process of technological change—nor does it enter into the highly detailed examination of particular cases of technological change and employment effects which have been the characteristic of a number of other studies. In adopting this particular standpoint, we have thought it desirable to produce a study which is at once capable of affording some basis for comparison between different companies and different industries during the implementation of technological change, and yet not so wide as to lose contact with the practical problems of specific instances of changing technology.

The main feature of the approach adopted was the development of a number of case studies in each of the three industries. These built up, from interviews and study of company records, an account of particular changes in technology at company level and the associated problems of planning, implementation and adjustment. The case studies were, as far as possible, selected so as to be representative of the types of technological change occurring in the industries, and this selection had therefore to be based on a preliminary investigation of each industry's technology. In this initial process we were greatly indebted to a number of people knowledgeable about the particular industry, especially in employers' or trade associations. The case studies, inevitably, obliged us to rely on assistance from company managements, and in the vast majority of cases we were given the utmost consideration and help. It may however be salutary to indicate two kinds of difficulty which we did encounter.

First, some cases which we would have wished very much to consider were in the event non-starters, since the companies concerned

7

were unable to see their way to giving co-operation. This was unfortunate from our point of view but understandable from that of the companies. There is no doubt that, both during and after the introduction of a technical change, management time is at even more of a premium than usual. Secondly, the case study approach depends to a great extent on the full co-operation of the companies involved. In a few instances, after work had begun, it was found that the amount of information available, or the amount of time available for discussion, was much less than we had hoped. For example, in the case of steel, the preparations for re-nationalization, which was only a vague possibility when we began our study, made for considerable difficulties. These points are not made to criticize the companies concerned, since management's prior commitment is obviously to the company's affairs. Nevertheless, we feel it desirable to point out this difficulty to others who may be tempted to adopt a similar method of approach. Despite this reservation, we are happy to record our sincere thanks to those companies, individuals and organizations who did participate in the inquiry, and to convey our special gratitude to a number of people whose interest, help and hospitality went far beyond the bounds of what might be regarded as conventional duty.

For reasons which are more fully explained in Chapter 1, we have used the case study information as a basis for a more rounded discussion of the experience of technological change at company and industry level. Case studies can be extremely useful for certain purposes, but in this instance we took the view that the case-study method should be used as a means to an end, rather than the end itself. It is hoped that this attempt to provide a balanced view of the problems typically arising during the process of technological change will provide a useful complement to other studies which have focused exclusively on individual cases.

In producing this book, all three authors have played a full and equal part. For the record, however, it should be said that each author undertook the major responsibility for a single industry. The actual allocation was: printing (D. B.), steel (L. C. H.) and chemicals (G. L. R.). In addition, Chapter 11 was undertaken by D. B., Chapter 12 by G. L. R. and Chapters 1 and 13 by L. C. H. Inevitably the study as a whole involved a continuing process of discussion and consultation among the authors.

It remains for us to make acknowledgement in a number of directions. First, we must acknowledge with gratitude the grant made to us by the Department of Scientific and Industrial Research (later the Social Scientific Research Council) which made this study possible. Secondly, we are greatly indebted to Professor D. J. Robertson for his advice throughout the study and particularly for assistance in the early stages, and to Mr D. I. MacKay who made a number of important criticisms and constructive comments during the final drafting stage. We have to repeat our thanks to those organizations, companies and individuals outside the University of Glasgow who contributed to the case studies and to our awareness of the industrial background. We are grateful also to the Department of Employment and Productivity and to the Ministry of Technology for making available to us certain statistics relevant to the study. And finally, we owe thanks to the secretarial staff of the Department of Social and Economic Research in the University of Glasgow, many of whom at one time or another took part in the work: especial thanks, however, goes to Mrs D. Ryder and to Miss E. Fairgrieve, who bore the main burden of typing successive drafts. Needless to say, any remaining errors of commission or omission are the responsibility of the three authors alone.

Department of Social   D. B.
and Economic Research   L. C. H.
University of Glasgow   G. L. R.

June 1969

CONTENTS

CONTENTS

PART ONE
INTRODUCTION

CHAPTER 1

AN INTRODUCTION
TO THE STUDY

I. TECHNOLOGICAL CHANGE AND MANPOWER POLICY

In recent years there has been a considerable extension of interest in manpower questions by those concerned with public policy. With the achievement of consistently high levels of employment and the emergence of an apparent shortage of labour in the United Kingdom, manpower policy is no longer concentrated upon eliminating the waste associated with large-scale unemployment. Rather it is concerned with improving the efficiency with which the available employed labour force is used, and with minimizing or eliminating pockets of unemployment which have not shared in the generally high level of economic activity.

The fundamental aim of such a manpower policy must be to improve the working of the labour market. If the economy is to grow and its structure to develop, it will be necessary for some workers to move from jobs for which demand is declining to those in which it is expanding. One cause of shifts in labour demand is changes in the structure of demand for goods and services. Quite apart from cyclical fluctuations which affect most industries, changes in incomes, relative prices and tastes bring about major alterations in the structure of demand which have their reflection in the labour market, where the need for a redistribution of labour between industries, occupations and geographical areas becomes apparent. In addition, industries are able in varying degrees to make use of scientific and engineering advances, making possible the production of new goods and offering alternative means of producing existing products. These too have their effect on the labour market, for changes in the technology of production are generally associated with some change in labour requirements per unit of output. Employment levels need not be adversely affected, though in practice this will often be the case, and it is also common for the occupational structure of the demand for labour to undergo change for this reason.

These changes in the demand for labour require equivalent

17

alterations in labour supply. In part, these may be achieved by not replacing those who retire from declining sectors, while new entrants to the labour force are attracted to the expanding sectors of employment. But where industries or occupations are experiencing rapid changes in labour demand, movements into and out of the labour force may be inadequate to cope with the required restructuring of employment, and increasing reliance will have to be placed upon mobility within the labour force. While many such movements will be accomplished voluntarily and without difficulty, adjustment in other cases may prove troublesome.

The development of national manpower policy, in this country as elsewhere, has been designed to improve the adjustment mechanisms in sectors of the labour market where normal mechanisms are in danger of proving ineffective for one reason or another. The extent to which technological change, by creating changes in demand for labour, falls into this category depends on the character of the change in technology itself, and on the rate of its introduction into the economy. In macro-economic terms, the possibility exists that technological change, in company with other forms of economic change affecting the structure of labour demand, may be biased towards the relative saving of labour,[1] thereby giving rise to the possibility of labour surpluses, resulting in unemployment on a large scale. This is of course a well-established topic in the literature of economics, and one which has been revived in recent years by the wave of change set up by the introduction, in many branches of industry and commerce, of computer technology and a range of innovations incorporating automatic controls. Individual applications of this kind of technology almost invariably produce savings in unit labour requirements and, when applied across a wide range of industries, inevitably give rise to speculation about an enhanced rate of displacement of labour and the prospect of widespread unemployment. The available historical evidence does not support the proposition that even intensive applications of particular types of technological change will generate large-scale unemployment, though in the course of the debate in the United States in the early 1960s,

---

[1] Econometric studies suggest that such biases do occur in practice. For example, in a study which measured the contribution of output growth, relative prices, neutral technology and non-neutral technological change to total changes in employment in a major sector of the us economy, Murray Brown found a bias towards labour saving during the period 1938-58: see *Theory and Measurement of Technological Change*, Cambridge University Press, Cambridge, 1966. See also M. Brown and J. S. de Cani, 'Technological Changes in the United States, 1950–1960,' *Productivity Measurement Review*, No. 29.

18

about the causes of high unemployment, the view was advanced that technological change had an important influence.[1]

It should, however, be recognized that quite apart from the general effects on labour demand and employment, there may be important biases in the utilization or saving of particular types of labour, which though they may have little impact on total employment, could seriously affect the employment prospects of some occupational and industrial groups of employees. What is currently recognized as a general trend in most developed economies towards extended use of administrative, technical and clerical manpower, and away from manual workers, especially the unskilled, could at least in part be explicable in terms of a bias in the direction of technological change. While technological change will tend to be guided by the need to economize on scarcer, high-cost factors, including certain types of labour, technology may be pushed to a stage where even a substantial reduction in the supply price of the high-cost factor would not be effective in promoting its immediate re-employment. For the time being, technological changes are not reversible and, in the short run at any rate, unemployment may result.

Clearly, then, the problems of labour market adjustment that may be generated by technological change are a justifiable concern of manpower policy. A change in technology may well add to the burden imposed on the adjustment mechanisms of the labour market by other forms of economic change and, furthermore, it may by its nature force mobility on workers who are unprepared for movement and thereby give rise to special problems. This latter point is important.

A great deal of the job changing that takes place is of course voluntary, where the worker leaves one job of his own accord and at a time of his own choosing, with the intention of finding other employment which may in fact already have been arranged. Even in the case of voluntary leaving, however, it appears to be fairly common for the worker to have no other job previously arranged.[2] But where job changing is involuntary, it is almost certain that the worker will have no alternative employment arranged at the time he receives notice of the termination of employment. A great deal then depends on the characteristics (skill and age particularly) of the

1 For a good statement of this view, see C. C. Killingsworth, 'Full Employment and the New Economics', *Scottish Journal of Political Economy*, February 1969 (and references therein).
2 For further discussion and references, see L. C. Hunter and G. L. Reid, *Urban Worker Mobility*, , Paris, OECD 1968, pp. 115–16.

individual, and on the employment situation in the relevant local labour market or in the economy at large.

This kind of circumstance will arise for reasons other than technological change. In a recession, jobs will disappear unexpectedly, though there may be a presumption that most, if not all of them, will reappear subsequently. In so far as they do re-emerge, these workers represent a temporary unemployment problem, quite different from those who are made redundant through structural or technological changes which may preclude the reappearance of jobs in their previous form. This is not to deny the seriousness of recessionary unemployment, since the 'temporary' period for some workers may be a matter of several months, and the number of workers that may be affected in this way in a short space of time will be great in relation to those made redundant (probably over a greater period of time) by any specific technological or structural change. There is also the fact that workers made redundant in a general recession are likely to find alternative employment hard to find, since most industries will tend to be in much the same position, whereas workers disemployed by technological change may find that current employment conditions are more favourable to re-employment—unless the technological displacement happens to coincide with recession.

Nevertheless, technological and structural changes present a very special kind of problem when they permanently eliminate a group of jobs, for in these cases mobility is forced upon the worker. How then does the mobility process work in such cases?

In the first place, the involuntary leaver does not have the same possibility of arranging the timing of his move. Even where relatively long notice of redundancy is given, the circumstances of the market may limit the relevant job opportunities to a very great degree, and may oblige the worker to consider jobs which carry lower pay and status. Secondly, the normal redundancy due to technological change is likely to hit both those workers who were in any case potentially mobile and those who, because of age, lack of skill or other reasons, were unlikely to consider voluntary transfer. In fact it may, if anything, discriminate against the latter since they are less likely to be adaptable to new jobs created by the technological change, and therefore less likely to be retained by the firm introducing the change. Thirdly, even though jobs with favourable wage and employment conditions are available, they may be quite unsuitable for the types of labour being made redundant. Fourthly, there may be serious institutional barriers to mobility.

If these are not to become major stumbling blocks both to the introduction of technological change and to the efficient operation of the labour market, policies have to be devised to deal with them at two levels. First, there is the problem for the company. To the extent that the introduction of a technological change involves a change in the demand for labour in the firm, policies will have to be developed to secure an adequate labour supply in both quantitative and qualitative terms, and if necessary to achieve a smooth transition to a new lower equilibrium level of employment. In the short run this will involve establishing procedures to deal with particular cases on an *ad hoc* basis, and may involve drawing up programmes for redeployment of workers within the company, providing training facilities for new labour and retraining for existing workers, and developing schemes to compensate redundant workers. It is important here to distinguish between job redundancy and worker redundancy: the former implies the disappearance of a job, the latter the failure or inability of the employing firm to offer new employment to the worker whose job has become redundant. In practice, as we shall see in the industry studies, firms have often gone to considerable lengths to avoid worker redundancy and to use the internal labour market of the company to ease the problem of re-employment, rather than impose an additional burden on the external market. In many cases, the problem may only be seen by the firm as a short-term difficulty; but in an industry where the experience of technological change is continuing, there may be a need to evolve a broader framework of policy, embracing continuous manpower forecasting for key groups of workers, a more comprehensive personnel policy, and a consistent and recognized procedure for handling the industrial relations problems of technological change.

Secondly, there are the problems for the economy. Up to a point, the size of the problem for the economy will be in inverse proportion to the success of the companies in dealing with their own problems of labour force adjustment. If companies can avoid large-scale displacement of labour into the market at large by careful planning, by the use of retraining and internal redeployment, and by expanding output to take up the employment loss embodied in the change itself, the country's labour markets will experience no great additional burden in the form of a flow of technologically displaced labour seeking new employment. However, even if companies in general were able to cope successfully with the problems of changes in technology, there is no certainty that the labour market as a whole will exhibit the necessary degree of flexibility to avoid misallocation and under-utilization of labour. But from the viewpoint of the national economy this is clearly an important question, and one for which manpower policy must show a close concern. Furthermore, since there is every likelihood that the treatment of technological change and its employment effects will vary in quality both between and within industries, it would seem desirable that public policy should be able to identify areas in which potential problems exist and to intervene directly or indirectly to secure an improvement in the approach of companies to the problems of adjustment. In short, national manpower policy has an interest both in the ability of labour market mechanisms to cope with the broad changes in labour utilization that are taking place, and in the efficiency of the more detailed adjustment practices and procedures adopted by individual companies.

Here again, there are both short-term and long-term aspects. In the short run it is possible for public policy to improve provision of income maintenance while workers are unemployed and seeking new work. Measures such as this, and the introduction of redundancy compensation, as well as easing possible hardships, may also serve an important function in the adjustment process by making it possible for displaced workers to take more time in looking for new jobs, rather than being compelled to take the first available job, which may be unsuitable and simply give rise to unnecessary movement subsequently. Still in the short run, the provision of facilities, for retraining workers with skills that are increasingly difficult to employ because of technological change, may also make a contribution to the adjustment process. In the longer run, the problems are rather different, being related more to the adaptation of the labour market system as a whole to make it more efficient in dealing with adjustment and change. Included in this will be measures to improve and expand manpower forecasting, to develop the educational and training needs of the changing economy, and to encourage the evolution of industrial relations attitudes and procedures suited to the types of change being experienced.

More will be said about these aspects in the final chapters of this volume, particularly in Chapter 13 where we consider the development of manpower policy in Britain, with special reference to its capacity for dealing with the labour problems generated by technological change. The major part of the book, however, is concerned with the trends of technological change in three selected industries, and with the labour force effects and adjustment problems that have arisen in practice. As we shall see subsequently, a great deal has been done in these industries to cope with the employment consequences of technological change at company level, and in particular cases studied in some detail there is no evidence of major problems of maladjustment. This is not, of course, to say that such problems do not arise. One of the difficulties of studying technological change is in tracing the employment effects not only in the companies and the industry in which it occurs, but in other companies and even other industries which may be adversely affected by a change in the market or competitive position resulting from technological change elsewhere. When this happens, it is much less possible for solutions to the adjustment problem to be sought internally and greater reliance has to be placed on external means of redeployment, which are of course more likely to result in unemployment, financial hardship and the need for assistance from public agencies.

Enough has now been said to indicate the area of our interest and its importance. There remains, however, a question about the appropriate method of approach to this set of problems, and it is to this that we now turn.

II. THE APPROACH TO THE STUDY

The main objective of this study, taken as a whole, is to examine some of the problems of technological change as they have emerged in different industries and different firms within these industries, with special attention to the employment and related labour issues of changing technology. There have of course been numerous recent studies which have contributed to our knowledge of these problems, especially on certain aspects of the adjustment process. A substantial literature now exists on the redundancy practices adopted by managements and on the effects of redundancy on workers affected by it. Sociological studies have been carried out on the effect of technological change on attitudes to new jobs, on changes in supervision arrangements, on employment and workplace status. Some attention has been paid to the implications of technological change for the role and structure of management, and also to the reaction of trade unions and work groups to the problems created by technological progress. But in the British context little has been done from an economic standpoint on the effects of specific technological changes in different industries, taking into account the nature of the changes in technology, their overall implications for the level and structure of employment, and the problems arising for the industries and their constituent companies in the context of particular labour market and industrial relations situations. This is the concern of the present study.

There are several reasons why this approach seems to be a useful one. In the first place, technological change is not a homogeneous process, but one which involves the application of widely varying types of scientific and technical knowledge to production problems. It is only to be expected, then, that the effects of changing technology on employment will be equally diverse, with some kinds of change producing little or no consequences for labour and others exerting very substantial effects indeed. It follows that we cannot proceed from an observation that technological change is taking place at a relatively high rate to the conclusion that proportionally large changes will be occurring in the employment sector. Secondly, the emergence and adoption of a new production technique, which in its application may have substantial labour force effects, need not be associated with widespread displacement of labour. Even though a new technique may be devised, capable of producing the same output with half the labour of the old, the extent of the employment effects will depend on the rate of introduction of the technique, the investment situation of the industry and its constituent firms, and the nature of the market for the product. Thirdly, changes in technology produce a bargaining situation in which wages, conditions of employment, worker selection and job structure come up for reconsideration, and the bargaining process itself may have a modifying influence on the purely technical potential of the innovation. For these reasons it has seemed to us desirable not just to consider the adjustment problems consequent upon technological change in isolation, but to regard them as being heavily influenced by the character of the change itself and by the industrial environment within which the change is implemented.

Because of this approach to the problems of technological change, the scope of the present study is wider than many others which have concerned themselves with the experience of particular companies or plants. This has meant that, for example, the detailed content of policies for adjustment adopted by individual managements has not been investigated as fully as in other inquiries. On the other hand, an important advantage of the present approach is that it makes it possible to set the broad outlines of management policy and union and worker reaction within an industrial and labour-market context which we regard as an essential element in the understanding of the process of technological change. Thus we have spent some time, in the following chapters, setting out the main features of the industry and describing the nature of the changes in technology which have been typical of recent experience. Some attempt has also been made to indicate the factors which have influenced the speed with which the industries have absorbed new technical possibilities. The central question, however, remains the employment and adjustment problems of technological change.

Before proceeding further, we should be more explicit about the definition of technological change used in this book. It is customary in economics to define a technique of production as a particular combination of labour and capital to produce a specified good. Labour in this context is defined broadly to include not only manual labour but also management and supervision services, and capital is again used in a wide sense to include materials, capital equipment and land or spatial requirements for the production process. A change in technique is thus a change in the combination of labour and capital inputs, which might take a variety of forms, ranging from minor adjustments in the use of labour to major changes in the type of capital equipment employed. Such a broad definition of technical change gives rise to practical problems. For example, it would include small changes in the use of labour which may have little to do with the actual technical processes of production and which may well be regarded by businessmen as administrative changes of a day-to-day character rather than as true examples of technical change. It would also presumably include the change in capital/labour ratios brought about by productivity bargaining, as a result of which jobs may be differently organized but the physical capital remains unchanged. In view of the present purpose it may seem advisable to concentrate on changes which actually originate in alterations to the capital equipment used in production. But this also raises difficulties. It will not always be the case that a change in the productive plant will involve a concurrent change in the factor inputs. For example, changes in the application of labour may be delayed or distributed over time, following a change in capital equipment, so that the full effect on labour demand is not always immediately apparent, while from the converse standpoint it may be difficult to determine whether observed changes in employment are due to some previous change in production technology or to some other cause. Despite this difficulty, which we encounter in practice in later chapters, we will concentrate here mainly on the labour implications of changes originating in the capital equipment, and to provide a continual reminder of this rather special usage we will refer to this as *technological* change, in preference to the more general concept of technical change as defined earlier in this paragraph.

Information on the problems of adjusting to changing techniques of production was built up on the basis of case studies in three industries: printing, steel and chemicals. It was decided, however, that the case study material should be used as a basis for discussing the problems of technological change, set within particular industrial contexts for the reasons just given, rather than simply presented as a series of unrelated studies. In each industry, study was made of eight to ten instances of technological change, so far as possible selected to represent different problems, but inevitably falling short of a fully representative sample of the industries in terms of company size, product structure and geographical location. The selection was made by a variety of means, but mainly through contacts with trade and employers' associations. Most, but not all of the firms invited to help in the study did so, but it is probable that there was an element of bias because firms prepared to co-operate in a study of this sort may be those with good records in these matters: those who got into difficulties may have been less willing to co-operate. But in fact several cases were at firms where quite serious technical or industrial relations problems have arisen. The detailed examination of cases of technological change in particular firms was supplemented by numerous discussions with other firms, trade union representatives, employers' associations and others with practical knowledge of the industries concerned— the intention being to get an accurate impression of the main trends in technology and of the kinds of problem arising.

A more fully representative sample of experience could have been achieved by concentrating attention on a single industry. We took the view, however, that more value could be had from a comparative approach. It is true that different companies, even in a single industry, will encounter different problems while introducing changes in technology. The experience of different industries may be even more diverse. However, it is arguable that there may nevertheless be common problems shared by different companies and industries, and that the comparative approach would show up the common and the peculiar elements.

The reasoning for the particular selection of industries was as follows. The printing industry represents a traditional craft-based industry which now has available to it a number of radically new techniques, partly based on research and development work in other industries, notably electronics and chemicals. It was thought that the impact of these new techniques, coming as they did after a long period of relatively static technology, would raise important issues and afford a contrast to industries with an experience of more continuous technological progress. It was felt, too, that the presence in the industry of strong craft unions, which are well known for their strong views on job property rights, might prove to be an important influence on the rate and effects of technological change in so far as it threatened to cut down established craft boundaries.

The feature of the chemical industry which seemed most interesting was that it is one of the most advanced scientifically and technologically. The fact that its output had grown rapidly in recent years, while the level of employment had remained almost static, suggested that developments in this science-based industry were making substantial reductions in unit labour requirements. It also seemed that there had been a shift in the structure of employment with relatively fewer process operators but rather more maintenance and technical workers. In view of the rapid technological changes which were taking place, involving changes in processes, raw materials, location and size of plants, it seemed that a study of the labour force consequences of these would be very revealing and the industry was then a strong candidate for inclusion.

Finally, the steel industry was included because it afforded ground for contrast and comparison with the other two industries. Although, like chemicals, it is a very capital-intensive industry, its growth record in terms of output has been much slower, and by 1964 a phase of over-capacity had been attained. The industry was at this time firmly embarked on a stage in its development during which advantage was being taken of a number of technical advances which seemed likely to bring about considerable changes in the structure of the industry. Furthermore, the trade union position was very different from that of the two other industries, and this too was an important factor in its selection. Since the study began, of course, the major part of the industry has been re-nationalized, but the case-study work was largely undertaken prior to this event.

A number of interesting comparisons became possible as a result of this selection. For example, the influence of the trade unions on employment matters varies considerably from industry to industry and significant contrasts in approach are to be found. From another point of view our industries are untypical of British industry as a whole, as they are all growing, with chemicals and printing experiencing rates of growth which are above the average for the

27

LABOUR PROBLEMS OF TECHNOLOGICAL CHANGE

economy as a whole. This means that the employment consequences of technological change have probably been mitigated to some extent by growing product demand, in contrast to other industries such as railways where greater productivity has exacerbated employment problems already present as a result of declining product demand.

The remainder of the book then takes the following shape. We begin with the printing industry, describing in Chapter 2 the structure of the industry and its industrial relations background, proceeding in Chapter 3 to a survey of the trends in technology in printing and their implications for employment, and considering in Chapter 4 the adjustment problems encountered and the methods used to solve them. The same procedure is followed in turn for the steel and chemical industries. Chapters 11 to 13 move away from the straight industry by industry approach, and take up particular issues which seemed to us in the course of the study to merit special attention. Chapter 11 involves a closer examination of the means by which standards of manning for new capital equipment are developed, and goes on to relate this to the important problems of under-employment of labour. Chapter 12, using evidence from our own studies and elsewhere, provides a general discussion of redundancy, resettlement and retraining, and their role as alternative courses of action in labour force adjustment to technological change. Finally, in Chapter 13, we return to the role of public policy in relation to the general labour market problems of technological change outlined in the first half of this chapter. This involves an identification of the problems which seem less likely to be capable of satisfactory solution by action at company level, and an evaluation of the way in which national manpower policy has been evolving in recent years towards an acceptance of responsibility for the residual difficulties of a labour market system faced with continuing and large-scale changes in technology.

PART TWO

THE PRINTING INDUSTRY

CHAPTER 2

THE STRUCTURE OF THE INDUSTRY

I. INTRODUCTION

In subsequent chapters we shall be discussing the main technological changes which have been introduced into the printing industry, and in particular the effect which they have had on employment in a number of specific cases. An understanding of the industry's experience of technological change depends, however, on viewing the case study material in the context of the industry as a whole, and it is the purpose of this chapter to provide a background to the discussion which follows. In Section II we therefore examine certain aspects of the newspaper and general printing industries, such as the size of firms, the structure of costs and the sources of revenue. This leads to a brief discussion of output trends, as these have been an important stimulus to the introduction of new techniques. Section III begins by examining the main characteristics of the industry's labour force, and then goes on to consider the main features of the industrial relations system in printing which are relevant to the later discussion.

II. COSTS, REVENUE AND OUTPUT

(i) *Definition of the Industry*

The 1958 Standard Industrial Classification divides the printing industry into two Minimum List Headings, as follows:

MLH 486 Printing and Publishing of Newspapers and Periodicals.
MLH 489 General Printing, Publishing, Bookbinding, Engraving, etc.

In this study we are concerned with all the printing and ancillary functions included under these two headings. While this division must be adhered to when using official statistics, the distinction is, in fact, somewhat blurred. Conditions of production in the firms producing national newspapers or mass-circulation periodicals

31

differ substantially from most other printing operations, and many small periodical and newspaper printers have more in common with general printers than with the large newspapers. The general printing industry includes the production of most other forms of printed material, such as books, packaging, stationery and miscellaneous publications. It also includes for statistical purposes the numerous, generally small, 'trade houses' serving the industry. These are firms who provide specialist services, such as typesetting or process engraving, to printers not possessing their own facilities.

(ii) *Newspapers and Periodicals*
(a) *Size structure*. The 1963 Census of Production showed that there were some 700 firms in this part of the printing industry. Of these, 17 firms with more than 1,000 employees accounted for nearly 68 per cent of the output of the industry. These large firms were naturally those engaged in the production of national daily and Sunday newspapers and of the mass-circulation periodicals. Since 1963, there have been several amalgamations in the industry and the number of separate large firms is now considerably reduced. In contrast, half of the firms employed fewer than 25 people. Some of these were small publishing firms without printing facilities, but the majority were the publishers and printers of small provincial newspapers which usually appear only once or twice a week.
(b) *Costs*. It is not the intention here to enter into a detailed discussion of the economics of newspapers and periodicals,[1] but there are some aspects of the industry's cost and revenue structure which are relevant to our subsequent discussion of technological change. Newspaper costs are shown by five categories in Table 2.1. It can be seen that the largest single item is newsprint and ink, the prices of which are largely beyond the control of the newspaper industry and which, incidentally, have risen more slowly than other costs since 1957. Editorial costs (i.e. the cost of collecting and editing news and features prior to printing) have been rising rapidly, probably reflecting an attempt to retain or improve the topicality and quality of newspapers in the face of competition from radio and television. Similarly, the rise in distribution and circulation costs is partly due to more extensive promotional activity as well as to

[1] A very full account is contained in the Royal Commission on the Press, *Report* (Cmnd. 1811, HMSO: London, 1962). This Royal Commission produced 6 volumes of Documentary Evidence and 3 of Oral Evidence, under the collective Command number 1812, as well as the *Report*. Future references to this source will be in the form, e.g. Royal Commission, op. cit., *Oral II*, Q.9645.

TABLE 2.1

Costs of National Newspaper Production by Category;
total and percentage, 1957 and 1965

Type of Cost	1957		1965		Increase 1957–65 (1957=100)
	Amount (£m)	% of total	Amount (£m)	% of total	
Editorial	10·1	13·1	22·0	14·6	218
Distribution and circulation	8·9	11·5	16·8	11·2	189
Newsprint and ink	30·9	40·1	47·3	31·4	153
Production Wages	17·7	23·0	42·4	28·2	240
Other	9·5	12·3	22·0	14·6	232
Total	77·1	100·0	150·5	100·0	208

Source: Economist Intelligence Unit: *The National Newspaper Industry, A Survey* (London, 1966), Part II, Table 30.

unavoidable increases in freight charges. But the item with the greatest percentage increase, which is also the second largest item in the total, is production wages. Further, production wages are directly under management control and the justification which can be made for at least part of the rise in editorial and circulation costs, namely the need to improve quality relative to that of competitors, does not apply. *A priori* one would expect this to constitute a strong stimulus to national newspapers to introduce labour-saving techniques.

To what extent may newspapers be able to stem this rise in production labour costs, by pooling production facilities to secure economies of scale? We must distinguish between those costs which do not vary with the number of copies of a newspaper produced and those which do. An important item in the fixed costs are the editorial costs of collecting and processing the news and other material up to the point at which it goes to the compositor. These do not vary with the circulation of the paper, at least in the short run. Furthermore, as the survey by the Economist Intelligence Unit showed, a national newspaper requires much the same amount of editorial expenditure whatever its circulation; for example, in 1965 this cost accounted for 22 per cent of the total costs of the *Daily Sketch*, but only for 11 per cent of the costs of the *Daily Mirror*, a comparable type of newspaper but with a much larger circulation.[1] Similarly, the costs of preparing the printing surface do not vary

[1] Economist Intelligence Unit: *The National Newspaper Industry, A Survey* (London, 1966), Part II, Table 15.

significantly with circulation. Consequently, these will be a higher proportion of total production costs in a newspaper with a small circulation than in a large one, and in 1965 they accounted for 31·3 per cent of the wage costs of a quality daily paper, but only 10·2 per cent in the case of a popular daily.[1]

In contrast to these fixed wage costs, those in the machine departments do vary with output. As output rises both the amount of machinery and the number of workers rise, and at the packing and dispatch stage the work load also rises with output. However, at high levels of output it may be worth putting in mechanized packing equipment to reduce wage costs, but this equipment is not so suitable for lower levels of output, where wage costs may therefore remain relatively high.

From this brief discussion it appears that the main area where wage costs could be reduced by merging production facilities is in the editorial and preparation departments, i.e. those departments where cost does not vary in direct proportion to output. The scope for greater editorial efficiency by sharing these costs amongst a number of newspapers is probably one of the main motives behind the various chains of provincial newspapers which are in existence, and was also a factor in the take-over of *The Times* by the Thomson Organization. Similarly, the concentration of typesetting facilities can produce useful economies without reducing the number of newspapers. This is particularly the case where a group of provincial newspapers under the same ownership may use certain common material such as stock exchange prices, racing results or editorial features. Equipment is available which enables the setting to be done once at a central location and signals are then transmitted by telegraph, causing type to be set in separate plants. We discuss this in more detail in Chapter 3.

If it can be arranged to print two or more newspapers which appear at different times on the same printing machinery, it is possible to achieve even greater economies. Labour costs would rise roughly in proportion to the level of output, but utilization of expensive plant would be much improved. We therefore frequently find evening and morning papers, or daily and Sunday ones, being produced in the same plant. A more recent development is where neighbouring provincial newspapers which appear on different days have begun to use the same printing machines: this has been

1 Ibid., Part II, Table 19. The examples chosen are *The Times* and the *Daily Mail*, and the figures in the text are obtained by adding together Compositors' wages, Foundry wages and Process wages in each case.

stimulated where one of them has invested in an expensive new printing technique and wishes to make full use of it.[1]

(c) *Revenue*. While a newspaper is naturally concerned with its level of costs, its ability to secure a sufficient revenue is generally more relevant to its profitability and survival.[2] The revenue of every newspaper derives partly from sales of copies, and partly from sales of advertising space. In 1965, 55 per cent of the total revenue of popular national newspapers came from sales of copies, and 45 per cent from advertising: for all other classes of national newspapers the advertising figure was higher than this, reaching 73 per cent for quality Sundays.[3] The 1965 figures are not available for provincial weeklies but the Royal Commission on the Press estimated that in 1960 some 79 per cent of their revenue came from advertising.[4] A newspaper is therefore competing for revenue on two fronts: with other newspapers and alternative news media for customers, and with other newspapers and quite different media for advertising expenditure. While there is not much competition for readers between different *types* of newspapers, there may be strong competition between papers in the same class for readers and also for advertising.

Rivalry between competing newspapers does not normally take the form of price competition. Demand for a particular class of newspaper appears to be price-inelastic, with only a slight loss of circulation following a general price increase.[5] The evidence is less clear, however, when we consider competition between newspapers in the same class. Many of the proprietors who gave evidence to the Royal Commission were of the opinion that a newspaper which increased its price, without the others doing likewise, would suffer a severe loss of circulation, implying that demand for a single newspaper would be elastic above the prevailing price level. But there is some evidence that demand for a newspaper is relatively unresponsive to price changes. For example, one newspaper which was trying to increase sales kept its price at 3d while its competitors raised their prices, but it did not attract many new readers. Similarly, when two Manchester evening papers were owned by the same company, an attempt was made to stimulate sales of the weaker one by raising the price of the stronger, but customers did not switch their purchases to the cheaper one in any substantial number.

[1] See Chapter 3.
[2] Cf. Royal Commission, op. cit., *Report*, Chapter 11, especially para. 253.
[3] Calculated from Economist Intelligence Unit, op. cit., Part II, Table 13.
[4] Royal Commission, op. cit., *Report*, page 23, Table D.   [5] Ibid., para. 179.

35

As demand for copies appears to be price-inelastic, competition for readers takes other forms. A publisher tries to increase the appeal of his newspaper to readers by making the content more attractive to them. These editorial considerations may have important consequences for production departments. If papers are in direct competition the topicality of their news becomes very important and strenuous efforts are made to shorten the production time of the paper to permit the inclusion of later items.[1] One effect of this may be to encourage the installation of new equipment which can be used to shorten the production periods, such as faster printing machines.

Competition to sell space to advertisers is also intense as the latter naturally seek to minimize their advertising costs per reader by concentrating their expenditure on what they consider to be the most effective newspapers. A fall in the number of readers may therefore mean not only a fall in circulation revenue but also in advertising. Because of the importance of advertising revenue to provincial newspapers there is an incentive to introduce new printing techniques which increase revenue from this source. This may come about because the appearance of the paper is brighter and more attractive to readers; more important, a process may have facilities which are highly valued by advertisers. As we shall see, the web-offset printing method has these advantages, which have stimulated its introduction into certain parts of the industry.

Similar considerations apply to the production of periodicals. Entry to the industry is easy, as most periodicals do not require the specialized equipment needed for newspapers and can be printed on general printing machines. Survival, of course, is another matter and depends on a publication securing sufficient advertising and sales revenue. In part this will be helped by high editorial standards but these, too, can be made more effective by use of the appropriate production techniques. This is particularly relevant in the case of trade and technical journals, which appear to benefit substantially in terms of their appeal to readers and advertisers if they contain large quantities of illustrations or colour. This can be achieved with any printing process, but, at the level of output normally achieved by trade and technical journals, web-offset is now generally considered the most efficient process. That this type of periodical is both expanding and profitable has been important in stimulating the introduction of the process by firms willing to expand in, or enter, this area of activity.

[1] Ibid., para. 182.

(iii) *General Printing*

(a) *Size structure.* In contrast to the newspapers and periodicals, large firms are relatively less important in general printing, as the 18 firms with over 1,000 employees accounted for less than 19 per cent of output in 1963, compared with 68 per cent in newspapers and periodicals. It follows that small firms are very important in this part of the industry, and more than 40 per cent of output in 1963 was by firms employing fewer than 100 workers. The continued existence of many small firms alongside some very large ones is explained partly by the nature of demand, and partly by technological factors. Most general printing is done on a jobbing basis; that is, it is done to a customer's specific requirements. Some requirements will be for very long runs of standardized products such as packaging, periodicals or paperbacks, permitting the use of specialized plant; many orders, however, will be for small quantities of miscellaneous printing work. The large heavily capitalized firms are not interested in these small orders, which small firms can therefore take up. They are helped by the almost indestructible nature of much printing equipment and they can therefore operate old plant indefinitely, also having access to a ready supply of second-hand equipment as larger firms re-equip.

On the other hand, it is often held that economic forces in the industry are operating to the disadvantage of the medium-sized firm between 25 and 300 employees. A recent study[1] of some fifty companies in the industry showed that the return on capital had declined from 13·4 per cent in 1961 to 9·7 per cent in 1966. Retained earnings are important in the finance of industrial investment, and lack of them may inhibit investment, particularly in the medium-sized firm. Even with the growth of office printing machinery, demand for the services of the small jobbing printer is expected to persist, while entry of new small firms to the industry at this level is easy and quite cheap. The large firms may be able to call on several sources of finance for investment in modern equipment to maintain or improve their profitability. But in between is the medium-sized, often private, company which for various reasons may be unable or unwilling to raise funds in the market. If its rate of return is around the average for the industry, it may not be able to generate a sufficient flow of funds to finance new investment, with the result that its equipment becomes progressively outdated and is less able to compete with bigger and more modern companies.

[1] Economic Development Committee for Newspapers, Printing and Publishing: *Financial Tables* (HMSO: London, 1967).

(b) *Costs*. The scope for economies of scale which can be achieved by a general printing firm is largely determined by the structure of its output. If it can concentrate on producing a limited range of standard items it will clearly operate much more efficiently in terms of unit costs than if there are substantial differences in the type of work done. Such concentration means that the company requires only a few of the wide range of machines of different types and sizes available, and as we saw earlier in newspapers, the unit cost of initial setting and preparation is reduced considerably if it can be spread over a large number of copies. One investigation[1] into book-printing costs illustrates this point. For the production of a 256-page book, fixed costs of composition, etc. were found to be £350. The variable cost per thousand copies consisted of £10 for printing, £30 for paper and £30 for binding. The effect of this was to reduce the cost per copy from 8s 5d if 1,000 were printed to 2s 1d if 10,000 were printed and 1s 6d if 50,000 were printed. Benefits also follow if the technical specifications such as the size of pages can be standardized, as this reduces costly adjustments to machinery between jobs.

(iv) *Output Trends*

Production figures for the printing industry are somewhat unsatisfactory being based on various input measures such as usage of paper or board, rather than on measures of output. However, it is evident from Table 2.2 that output measured in this way has been rising at a rate of about 5 per cent a year, which exceeds that of industrial production as a whole. Given the variety of printed products and the different influences determining output levels, it is worth examining the constituent elements a little more closely.

(a) *Newspapers*. There has been concern in recent years over the state of the newspaper industry following a number of closures. Table 2.3 suggests, however, that there is considerable variation in the experience of different types of publication. It is clear that there has been a substantial decline amongst provincial morning newspapers and amongst popular Sunday newspapers. However, the number of provincial evening newspapers has remained almost constant, and since 1966 several new ones have been started.[2] In addition, there has been a considerable growth in the sales of the national 'quality' newspapers, particularly on Sundays. In the latter

[1] C. Pratten and R. M. Dean, *The Economies of Large Scale Production in British Industry: An Introductory Study* (Cambridge University Press, Cambridge, 1965).
[2] See the *Newspaper Press Directory* 1968 (Benn Bros. London), p. 11.

TABLE 2.2

*Index of Industrial Production for Manufacturing Industries
and Printing and Publishing (1958=100)*

	1959	1960	1961	1962	1963	1964	1965	1966	1967
All manufacturing industries	106	115	114	115	119	128	134	136	134
Printing and Publishing	107	120	120	124	131	142	145	148	150

Note: The output figures for the printing and publishing industry are compiled from input sources such as the consumption of newsprint and other selected items of paper. This probably leads to an understatement of the output of the industry as, for example, more work is involved in printing something in four colours instead of one and there has been a considerable growth of colour printing in recent years.

Source: Information supplied by the Central Statistical Office.

TABLE 2.3

*Number of Newspapers, and Circulation, by Class,
1947, 1961 and 1966*

	1947		1961		1966[1]	
	Number of Publications	Total Copies (000's)	Number of Publications	Total Copies (000's)	Number of Publications	Total Copies (000's)
Morning Dailies:						
National quality	4	1,480[2]	4	1,878	4	2,062
National popular	7	14,280	6	13,934[3]	5	13,509
Provincial	24	2,574[2]	18	1,800	N.A.	N.A.
Sundays:						
National quality	2	952	3	2,370	3	2,877
National popular	8	24,287	5	22,166	5[4]	21,329
Evenings:						
London	3	3,500	2	2,203	2	1,904
Provincial	75	6,800	74	6,700	N.A.	N.A.
Weekly	1,307	N.A.	1,219	N.A.	N.A.	N.A.

Notes: 1. Figures based on first six months.
2. *The Guardian* has been counted as a national rather than a provincial newspaper.
3. Excluding one very small newspaper which had come into existence, with a circulation in 1961 of 23,000 copies.
4. The *Sunday Citizen* ceased publication later in 1966 reducing this to 4.

Sources: Royal Commission, op. cit., *Report*, Tables A, B, C and paragraph 10; and Economist Intelligence Unit, op. cit., Part II, Table 2.

case, not only have sales grown, but so has the size of newspapers. The EIU survey showed[1] that the number of pages sold in 1965 had increased by 111 per cent over the 1957 figure for quality dailies, by 30 per cent for popular dailies and by 269 per cent for quality Sunday newspapers. These figures reflect the growth of advertising expenditure in general, and 'financial' and 'prestige' advertising in particular. Growth in circulation and size severely strained the existing capacity of some newspapers, and as we shall see later, has been an important stimulant to the introduction of certain new techniques.

(b) *General printing and periodicals.* Table 2.4 shows the movement in output of certain sectors of the general printing and perio-

TABLE 2.4

Net Output of General Printing Industry 1954, 1958 and 1965 (Estimated); 1958 Prices

	Actual			Estimated	
	1954 (£m)	1958 (£m)	% annual change 1954–58	1965 (£m)	% annual change 1958–65
General printing including books	146·6	170·3	3·9	204·1	3·7
Periodicals	55·0	60·4	2·3	66·4	1·9
Cartons	17·3	19·8	3·5	22·6	2·7
Stationery	22·7	25·4	2·8	29·9	3·3
Books (inc. above)	13·5	19·1	9·1	23·3	4·1
Total general printing	241·6	275·9	3·4	323·0	3·2

Source: Census of Production, 1958, and British Federation of Master Printers, *Economic Study of the Printing Industry*, 1965.

dicals sector, and estimates for 1965. It is seen that book printing is the most rapidly growing sector of those shown, but constitutes only about 7 per cent of total general printing work. Since 1954, expenditure on books has been rising at almost 3 per cent annually in real terms.[2] On the other hand, periodical production has been growing relatively slowly, reflecting a decline in consumers' expenditure on this type of publication following the spread of television ownership. One type of periodical which has been growing,

[1] Economist Intelligence Unit, op. cit., Part II, Table 4.
[2] Cf. 'Paper and Board, Trends and Prospects', NIESR. *Economic Review*, May 1965, Appendix Table 19.

however, is the 'trade and technical' magazine for business interests. As we discuss later, this has been the specific stimulus to the introduction of web-offset in several firms. The demand for other printed products such as packaging has grown, as one might expect, roughly in line with the growth of total consumers' expenditure.

III. LABOUR FORCE AND INDUSTRIAL RELATIONS

(i) *Labour Force*

Although output has been increasing at an average rate of 5 per cent per annum, it can be calculated from Table 2.5 that the printing industry labour force has only risen from some 358,000 in 1959 to

TABLE 2.5

Estimated Numbers of Male and Female Employees
(Employed and Unemployed), May 1959 and June 1967

	1959		1967	
	Number (000's)	*Percentage of total*	*Number (000's)*	*Percentage of total*
MLH 486				
Males	100·9	78·5	110·8	76·3
Females	27·6	21·5	34·4	23·7
Total	128·5	100·0	145·2	100·0
MLH 489				
Males	142·6	62·0	165·6	63·3
Females	87·2	38·0	95·9	36·7
Total	229·8	100·0	261·5	100·0

Source: Ministry of Labour *Gazette*, February 1960 and March 1968.

nearly 407,000 in 1967. This amounts to 1·5 per cent per annum, on average, over the period. There is an interesting difference between the occupations of women in the two parts of the industry. In MLH 486, the newspaper and periodicals sector, the women are mainly in the Administrative, Technical and Clerical Classification, constituting only a small part of the operative labour force and this is almost entirely in the production of periodicals as distinct from newspapers. In general printing they provide nearly two-fifths of the operative labour force, but this proportion has tended to decline. This trend, quite contrary to that in many industries, is thought to reflect increasing mechanization of parts of the printing operation, especially at the binding and finishing stage. Certain types of machines are the prerogative of male workers, and as these become more common,

41

or previously separate operations are linked to them, the demand for female labour falls relative to that for men.

There is no fixed retirement age in the industry, and many male workers stay on past the age of 65: printing employs 4·3 per cent of all male employees, but 5·6 per cent of those aged 65 and over.[1] As we shall see, several firms have been able to bring about quite substantial reductions in their labour force when required, simply by persuading those over 65 to retire.

The most striking feature of the male labour force in general printing is the high proportion who are classified as skilled operatives, as shown in Table 2.6. This figure must be interpreted with care

TABLE 2.6

Analysis of the Printing Industry Labour Force,
by broad occupational groups May 1967

	Number in each occupational group as percentage of total				Total		
	Administra-tive, technical and clerical	Skilled	Semi-skilled	Others	Per-centage	Number	In Training[1]
MLH 486							
Males	36·6	45·1	2·0	16·3	100·0	107,340	4·7
Females	76·1	9·8	0·3	13·8	100·0	28,520	4·6
Total	44·9	37·7	1·6	15·8	100·0	135,860	4·6
MLH 489							
Males	22·4	65·4	1·2	11·0	100·0	150,020	10·9
Females	31·6	52·1	6·4	9·9	100·0	83,430	9·3
Total	25·7	60·6	3·1	10·6	100·0	233,450	10·3

Note: 1. The percentages for those in training are excluded from the total figures.

Source: Ministry of Labour *Gazette*, January 1968; Tables 24–26, pp. 36–40.

as it includes some workers whose skill is acquired by several years' experience such as machine assistants, who have not served apprenticeships and are not recognized as skilled within the industry. But even if these are excluded, about 45 per cent of the male labour force has served an apprenticeship. The proportion of skilled workers is lower in the newspaper and periodical industry, which also trains a much lower proportion of its skilled workers than general printing.

[1] Ministry of Labour *Gazette*, June 1967.

The occupational structure of the industry is shown in more detail in Table 2.7 which shows that the skilled part of the labour force is divided into a number of clearly defined occupations. This stems

TABLE 2.7

Principal Craft Workers in the Printing Industry, May 1967

Occupation	Male	Female
Compositors	46,240	560
Electrotypers and stereotypers	3,830	—
Letterpress machine minders	21,390	1,290
Bookbinders, cutters, etc.	13,070	15,990
Litho minders and printers in metal	5,060	140
Litho artists	4,150	300
Gravure machine minders	2,000	—
Gravure artists	2,900	—
Process engraving workers	2,750	60
Press telegraphists	700	10
Other skilled workers	9,440	2,670
Maintenance	7,070	230
Total Craft Labour Force	118,600	21,250

Source: Ministry of Labour *Gazette*, January 1968.

from the historical growth of the trade union organization in the industry, which divided the whole printing operation into a series of specialized and separate crafts, though in recent years this trend has been reversed with the amalgamation of printing unions. Consequently most of the remaining unions now contain several quite separate crafts: this, however, is an aspect of the mobility of labour in the industry and we shall defer discussion until Section III (iv) below.

(ii) *Trade Union Organization*

The industrial relations system in printing has a much greater influence on the course of technological change in the industry than is the case in either steel or chemicals. The structure of the trade unions in the industry is historically based and several new techniques cut across traditional boundaries. Furthermore, in some cases the unions have placed considerable emphasis on protecting their members' employment opportunities in the face of technological changes, while a strongly developed system of plant negotiations traditionally covers a range of production matters.

The trade union side of the industry has recently undergone a substantial change in appearance. At the time of the Royal Commission on the Press[1] in 1961, there were twelve printing trade unions in the industry who were members of the Printing and Kindred Trades Federation, but subsequent amalgamations have reduced the number to five with further mergers under consideration. The present unions and their membership are shown in Table 2.8. The amalgamation movement has had the effect of polarizing

TABLE 2.8

Printing Trade Union Membership, 1968

Trade Union	Membership
National Graphical Association	105,237
Scottish Typographical Association	6,985
Society of Lithographic Artists, Designers, Engravers and Process Workers	16,014
National Union of Journalists	20,271
Society of Graphical and Allied Trades	228,902

Source: Based on Trades Union Congress *Report*, 1968..

the unions in the industry into two large unions with a number of small unions still retaining their independence. The National Graphical Association (NGA) was formed from 6 of the 12 unions existing in 1962 and its membership is restricted to printing workers who have served an apprenticeship in the industry. The main occupations in which it recruits are composing and machine-minding although following an amalgamation in 1967 it also includes the stereotypers. Its members have traditionally been in charge of letterpress and gravure machines, but, following a merger in 1968 with the Amalgamated Society of Lithographic Printers, are now responsible for offset machines as well. Similar groups are covered by the Scottish Typographical Association (STA) in Scotland. By contrast to the closed nature of the NGA, the other large union, the Society of Graphical and Allied Trades (SOGAT) includes both craft and non-craft workers. It has absorbed four of the unions existing in 1955, and is now by far the largest union in the industry. Obviously the range of occupations which it covers is great, but the two which will be our main concern subsequently are the non-craft machine assistants on all types of printing machines, and those engaged in the finishing operations of the printing and publishing process such as cutting, binding and packing. Finally, the Society of Lithographic Artists, Designers, Engravers and Process Workers (SLADE) organizes workers who are engaged mainly in the preparatory work necessary for printing illustrations. All of the printing unions

[1] Royal Commission, op. cit., *Report*, Appendix XI.

44

are members of the Printing and Kindred Trades Federation (P & KTF) to which we return in Sub-section iv.

(iii) *Employer Organization*

There are two major employers' associations. The national daily and Sunday newspapers are all members of the Newspaper Publishers' Association (NPA).[1] While it performs the function of a trade association, its main work involves labour matters. Major negotiations are conducted by a Labour Committee reporting to the NPA Council, which has representatives from all member firms. In Scotland, the Scottish Daily Newspaper Society negotiates basic agreements applying to publishers of daily newspapers in Scotland, although these vary from small local dailies to those selling throughout Scotland. The Newspaper Society is an association of publishers and printers of provincial daily and weekly newspapers in England and Wales.

The second major association is the British Federation of Master Printers (BFMP), whose 4,000 members in the general printing industry employ about 80 per cent of the industry's labour force. National negotiations are conducted jointly with the Newspaper Society. The Federation is composed of 12 regional Alliances, including the Society of Master Printers of Scotland. The latter and the London Alliance are parties to the national agreements, but each holds some separate agreements with the unions on particular matters.

(iv) *Collective Bargaining*

The main negotiations on wages and conditions are conducted between the P & KTF and either the NPA or the BFMP and Newspaper Society acting together. Some agreements, such as the basic hours and holidays agreements and one laying down apprentices' wages, are made between the employers' associations and the P & KTF acting in its own right. On other matters such as minimum wage rates or the value of overtime and shift bonuses, the P & KTF simply co-ordinates the claims of the separate unions and negotiates on their behalf. Any agreement reached by the P & KTF in this way must be submitted by each union to its members for approval before ratification, and the agreements themselves are held by the individual unions with the employers' associations. The unions retain the right to negotiate independently. This they sometimes do, submitting claims concerning their own membership in addition to the general P & KTF claim.

[1] Until 1968 this body was called the Newspaper Proprietors' Association.

45

Although the national negotiating machinery in printing is well established, a feature of the industry is the strongly developed bargaining machinery at the plant or 'house' level. Each union has its network of local branches responsible for functions such as maintaining or extending the level of union membership, the operation of an 'employment exchange' for members, and some negotiating activities. In addition, members of each union belong to the 'chapel' of that union in their particular plant. They elect as their representative the Father of the Chapel (FOC), analogous to the shop steward in other industries. Although firms will usually be members of an employers' association, they remain free to conclude house agreements with unions in their plants provided only that the nationally agreed minimum rates and conditions are observed.

(a) *Wages.* Average earnings of manual workers in printing are amongst the highest in manufacturing industry. In October 1967, average earnings in newspapers were more than £29 a week and in general printing about £24 a week. The average for all manufacturing industry was just under £22 a week.[1] Although basic rates are higher than many industries, the main source of the difference lies in the difference between nationally negotiated rates and actual earnings. The minimum rates agreed nationally may be supplemented by numerous plant-based extra payments such as regular overtime working, merit money, timekeeping bonuses and the like. In an industry with increasing product demand, a slowly growing labour force and strongly developed traditions of plant negotiations, it seems almost inevitable that substantial wage drift will occur, and general printing illustrates this. Between 1960 and 1967, weekly wages rates rose by 33 per cent, while average weekly earnings rose by 47 per cent. The most likely explanation is that the established plant bargaining machinery has enabled unions to make the most of conditions of labour shortage in the industry. In order to secure and retain their labour force, firms have been prepared to pay merit and other house payments, or to work heavy overtime. There seems to have been little difficulty in passing on increased costs in higher prices.[2]

There is virtually no available information on earnings differentials within the industry, although Table 2.9 gives some idea of the range between craftsmen's and assistants' wages. Because of the small sample upon which these data are based it would be dangerous

1 Ministry of Labour *Gazette*, February 1968.
2 National Board for Prices and Incomes, Report No. 2, *Wages Costs & Prices in the Printing Industry* (Cmnd. 2750, HMSO: London, 1965), paras. 52 and 53.

to place too much weight on them, but they do suggest that the gap between the earnings of craftsmen and those of their assistants is

TABLE 2.9

Average Weekly Earnings of Certain Groups of Printing Workers, 1966

	Machine Minders (£s)	Machine Assistants (£s)
Letterpress:		
Sheet-fed	20–35	19–28
Web-fed	25–35	20–25
Offset-litho:		
Sheet-fed	28–50	21–31
Web-fed	27–45	18–25

Note: The information on which this table was based was obtained by the BFMP from a small sample of their members and should be treated with care.

Source: Court of Inquiry into the problems caused by the introduction of web-offset machines into the printing industry, *Report* (Cmnd. 3184, HMSO: London, 1967).

greater on web-fed litho machines than on other types of equipment, which may at least in part explain the disputes which have arisen over the manning of this type of equipment, and to which we refer in the later chapters.

(b) *Supply of labour*. The number of boys apprenticed to each craft is governed by a ratio of apprentices to craftsmen agreed between the BFMP and the union concerned. A firm can only

TABLE 2.10

Examples of Apprentice Ratio Calculation
(for Composing Room)

Number of Craftsmen Employed	1–4	5–8	9–14	15–20	21–26	27–34	35–50
Number of Apprentices Allowed	1	2	3	4	5	6	7

Note: The allowance continues above 50 employees at a rate of one apprentice to 16 craftsmen.

Source: *Graphical Journal*, March 1966.

employ apprentices in relation to the number of trained craftsmen it already employs. The formula varies slightly between unions but the principles of calculation are similar to those shown above for composing room apprentices. It is apparent from Table 2.10

47

that large firms are able to employ proportionately fewer apprentices than small firms; for example, a firm employing 40 compositors can train only seven apprentices, whereas a firm with four compositors can train one. The agreements also lay down that the length of apprenticeships shall be of five or six years duration, depending on the age at entry of the person involved.

Table 2.6 showed that newspapers train very few workers in relation to their needs, and this also applies to the number of apprentices whom they train. While apprentices in general printing (MLH 489) amounted to 13·5 per cent of male skilled operatives in May 1967, the comparable figure for the newspapers and periodicals industry was 6·5 per cent; and most of these would be accounted for by periodical printers who train apprentices in much the same way as general printers.[1] That national newspapers train so few is partly because most of the work is done at night, but the trade unions also claim that the training a newspaper would give in most departments would be too specialized, and that a person trained on a national newspaper would have difficulty in securing employment in the general printing industry. Where the craft is common to the latter (as in foundry work) some apprentices are trained. One newspaper has some Monotype composing machines, used in the general trade, as well as the usual newspaper Linotypes and therefore trains some Monotype operators. But most newspaper labour is drawn, fully trained, from the general printing trades: earnings are much higher than in the latter and no difficulties are experienced in securing a supply.

The entry of non-craft workers to the industry is controlled in a similar though less formal way. The male and female 'learners' which a firm in the general trade may employ are restricted in number to the customary level for that plant, for example 10 or 15. New learners can only be employed as existing ones complete their training. As with craft labour, a large firm may find that it is unable to meet all its requirements for labour from those it trains itself. In this event the union concerned will supply it with the necessary staff by drawing people from smaller firms, who will then proceed to train further supplies. Entry to the better-paying large firms thus tends to be through the smaller firms.

(c) *Manning standards*. The apprenticeship and other agreements outlined above clearly represent an effective control on the supply of labour to the industry. Collective agreements also influence the demand for labour in the industry, mainly by specifying the manning

[1] Ministry of Labour *Gazette*, January 1968; calculated from Table 24, page 37.

necessary on certain types of equipment, but also by agreements limiting a firm's right to declare redundancies, or where a firm may guarantee to provide a union's membership with a given number of jobs if required to do so. For certain equipment in general printing, the number of people who must be employed to operate it is specified in national agreements. For example, it is agreed that a large rotary gravure machine will be manned in accordance with the following scales[1]:

> up to 3 cylinders in use—2 minders
> 4 to 6 cylinders in use—3 minders
> 7 or 8 cylinders in use—4 minders.

Similarly, there is an agreement that when teletypesetting equipment is used the caster operator shall not be in charge of more than two casting units. The establishment of these national agreements is made possible by the standard nature of much of the equipment in the industry. Much preparatory printing and finishing equipment is purchased by a printer from a standard range, and, unless there is some peculiar feature in his plant, it will require much the same manning as in any other plant. This is in contrast to steel and chemicals, where many plants are built to unique specifications, and manning has to be worked out for each plant. When a printing firm instals new equipment not covered by a national agreement, a procedure is laid down which provides for the establishment of a joint inspection team composed of union and management representatives, who will establish a manning figure applicable to that installation and future ones. While a decision is being reached, the machines are operated for one week by the manning proposed by management and the next week by that proposed by the union, and so on until agreement is reached. In practice, agreement will often be reached between management and the chapel concerned, and such a house agreement would stand, irrespective of what the joint team might subsequently decide. Another aspect of manning is that certain machines in binderies are claimed exclusively by men, and women are not allowed to operate them.

(d) *Mobility and redundancy.* As technological changes occur in the industry and cut across traditional craft boundaries, it is important to assess the ease or difficulty with which a person displaced from one occupation can move to another, and the institutional factors helping or hindering this movement. The rigid structuring of the labour force which we discussed earlier has the effect that it is

[1] There would also be a complement of assistants on the machine.

generally very difficult for a person who has entered a particular craft to move to another one. In part this reflects genuine training or skill barriers but the reluctance of unions to permit entry by those who have not served the appropriate apprenticeship is more important. There was, however, an agreement between the NGA and the ASLP, before the two unions merged in 1968, for inter-union transfer and retraining in the event, but *only* in the event of a firm changing its printing process. But in the 10 years of the agreement, less than 100 transfers have actually occurred. A rather more active transfer and retraining between crafts is carried out within SLADE where the union has transferred a number of men from its declining process-engraving section to other expanding areas. This is a good example of movement taking place within a union, even though its membership is highly structured by occupation. But arrangements of this type are unusual, and most of the labour force is clearly and firmly stratified according to occupation.

The mobility of labour within the industry is not only regulated by the control exercised over movement between occupations. In general the agreement of the union at branch level is required before a man can move from one firm to another. The common practice, when an employer needs additional labour, is for him to inform the appropriate union branch, which will then attempt to fill the vacancy either from unemployed workers or more likely from a list of members who have indicated a desire to move from their present firm. Firms can of course advertise for workers, but the latter are usually required to inform their branch before moving. Often these communications are mere formalities and the degree of control varies, but SLADE, for example, tries to keep a strong measure of control over its members' movements.

Certain *ad hoc* redundancy agreements have been made in the industry. For example, from 1960 to 1963, an agreement existed between the NPA and the National Union of Printing, Bookbinding and Paper Workers to the effect that no redundancy of regular staff would be declared upon the introduction of mechanized tying equipment, which had substantial labour-saving potential. Further, a number of firms have an agreement with the Society of Lithographic Artists, Designers, Engravers and Process Workers to the effect that they guarantee a given level of employment to members of that union, in return for which the union guarantees to meet all the firms' demands for labour. The firm does not have to employ its 'guarantee' figure all the time, but if some members of the union were unemployed, the firm would be obliged to take them on up to its guarantee level. These and other agreements at national and plant level have sometimes been of particular importance in relation to the impact of technological change on employment, and will therefore be discussed more fully in the context of the case studies.

In this introductory chapter we have discussed many aspects of the printing industry which are relevant to our subsequent discussion of technological change. One of the most important is that the output of the industry has been growing more rapidly than that of industrial production as a whole. As we shall see below, this growth in demand has been an important stimulus to the introduction of new techniques, and at the same time has often mitigated their employment consequences, in that new plant has often served to prevent an increase in a firm's labour force, rather than to reduce it. We have also shown that competition, particularly in the newspaper and periodicals sector, often concentrates on product quality rather than price. This too has been instrumental in the adoption of certain techniques which, while not reducing costs, nonetheless improve the appearance of the finished article. The main points to arise from the outline of the industry's labour force are the large proportion of the employees who are skilled craftsmen, and the many specialized occupations into which the labour force is divided, both of which have implications for the adaptation of the labour force to new production methods. Finally, we have discussed a number of ways in which the industrial relations system in the industry may affect the outcome of technological change. In the printing industry there is an established tradition of bargaining not only over wage matters but also over manning standards and questions of mobility and redundancy. The policies of the trade unions towards technological change are therefore an important consideration in the printing industry and they consequently figure quite prominently in much of the subsequent discussion.

CHAPTER 3

THE EMPLOYMENT EFFECTS OF
TECHNOLOGICAL CHANGE

I. INTRODUCTION

In the present chapter we are primarily concerned with the presentation of the evidence arising from our case studies in the printing industry concerning the effect of technological change on employment. Naturally a complete assessment must include a consideration of such matters as output trends, the rate of adoption of new techniques, and the policies of the trade unions towards technological change. However, these and related questions, which place the case-study material in the context of the industry as a whole, are the subject of the next chapter. Here we are only concerned with the case-study material, and our discussion will be in three parts, each dealing with one stage of the production process.

Almost every aspect of the printing operation has been subject to some significant technological advance in recent years, although the extent to which these advances have been adopted by firms in the industry varies considerably. In reviewing their effect on employment in the present chapter, we shall consider several questions. The first concerns the effect of the new equipment on the amount of labour required per unit of output. Although this apparently simple question is complicated by several factors, it is possible to indicate in broad terms those new techniques which have significant labour-saving effects, and those which do not. Where these labour-saving effects are considerable we must then consider how the necessary quantitative adjustments to the labour force were achieved. But as well as changes in the size of the labour force, new techniques may lead to qualitative changes, for example in the type of skill required. Again we must then ask how the necessary adjustments were brought about. This question is particularly interesting in the printing industry because an unusually large proportion of this industry's labour force has traditionally been trained in highly specialized skills. Consequently, new processes which cut across the established boundaries between different skills might be expected to cause more than usually difficult adjustment problems.

It may be helpful to precede the discussion with a brief outline of what is involved at each stage of the production process, with the aid of Diagram 3.1 overleaf. The central and most easily understood part is the actual printing stage, when the inked printing surface and the paper are brought into contact. There are three basic processes, but the principles of each process can be applied in more than one way, so that there are seven different techniques. These are shown at the second level of the diagram. However, before the printing stage, several prior operations must be performed to set type, make illustrations, make up the page, and prepare the printing surface. These activities are known as the preparatory stages, and can be seen at the first level of the diagram. After material has been printed, a series of finishing operations are generally required to bind, pack and dispatch the product. The whole operation of producing a printed article proceeds in the order shown in the diagram. Though it might seem logical to consider each stage of production in the order in which it occurs, it is in fact easier to deal with printing first, and then go on to the preparatory and finishing stages. The subsequent sections therefore follow this arrangement.

II. THE PRINTING STAGE

(i) *The Alternative Techniques*

The three basic printing processes are letterpress, offset and gravure.[1] The principles of each process are embodied in equipment designed to print on paper which is either in sheets or in reels, giving six possible techniques. In addition, the technique known as flat-bed letterpress uses a flat printing surface as distinct from the cylindrical surface used by the others. Each of these seven techniques is, of course, available in different designs and capacities, further widening the range of choice available to a firm.

Letterpress is the traditional printing process, used until recently for all newspapers and a wide range of general printing. The principle is that the printing surface is raised relative to its surroundings, so that when an ink roller passes over the surface, ink is deposited only on the printing areas. The ink is in turn transferred to the paper when this is brought into contact with the printing surface. *Flat-bed*

---

[1] There are certain other experimental and 'hybrid' techniques, but these are generally at an early stage of development and it is not necessary to give an account of them in this context.

DIAGRAM 3.1

Stages of Production and Principal Alternative Techniques

Preparatory stage
    typesetting

illustrations

production of the
printing surface.

Printing stage

Finishing stage

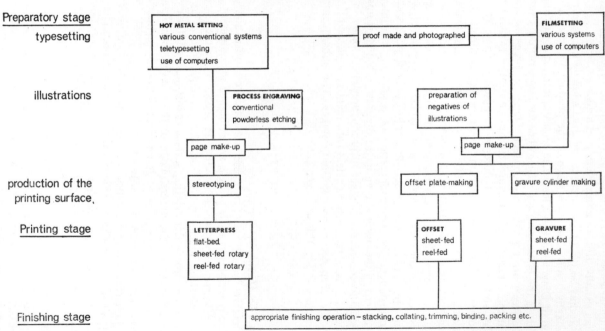

letterpress machines are the oldest form of printing equipment and, as the name implies, they use a flat printing surface which moves back and forth, printing one sheet in each complete movement. As half the operating time is spent by the printing surface returning to its original position, the output of these machines is relatively low, some 2,000 to 2,500 sheets an hour. Much faster speeds are obtained on the *reel-fed rotary letterpress* machines, and depending on the quality required they can print as many as 50,000 copies an hour. The principles upon which flat-bed and reel-fed rotary letterpress machines operate are of long usage, and the scope for further improvement in successive machines is limited. In contrast the *sheet-fed rotary letterpress* technique has recently been the subject of a more substantial technological advance. Although the equipment has been available for many years, an obstacle to its wide use was the difficulty of producing a satisfactory curved printing plate.[1] This problem was overcome in the late 1950s and the machines have been installed by a number of letterpress printers, either as additions to or replacements for existing flat-bed letterpress plant. These machines print about 6,000 copies an hour.[2]

Offset-lithography, usually abbreviated to offset, uses as a printing surface a smooth metal plate on which only the printing areas have been made receptive to ink. The plate is fitted round a cylinder on the printing machine, and as it revolves picks up ink on the printing areas. The ink is then transferred, usually via a second cylinder, on to the paper. The process originated in the nineteenth century and has for many years been used for cheap and lower quality printing work. But in recent years there have been major advances in the technology of offset printing.[3] Although there have been mechanical improvements to offset machines, the most important changes have been in the quality of the materials used. In particular the ink manufacturers have improved the quality of inks which are available, so that the process is now capable of a much higher quality of printing. The work of making the printing plates used to be a skilled and time-consuming job, but it has been greatly simplified and is now done easily in a few minutes. Lastly, improvements in the

[1] In newspapers the quality of impression is less important, and quite satisfactory methods have been evolved. The problem is greater where higher quality printing is required.
[2] See, for example, W. R. Durrant, 'Progress Report on Sheet-fed Rotary Letterpress', *British Printer*, April 1968.
[3] See, for example, Peter Robinson, 'Web-offset: Progress and Prospects', *Penrose Annual*, 1963, and Michael H. Bruno, 'Research on Paper for Printing', *Penrose Annual*, 1968.

strength and other features of paper for offset printing have facilitated the development of the reel-fed version known as web-offset, which was first used in the United Kingdom in 1957. These improvements have raised the quality of offset printing, and widened the range of work which it can produce more economically than other processes.

Finally, the gravure printing surface is a copper cylinder treated in such a way that the printing areas are lower than their surroundings. Ink flows into these areas and is drawn on to the paper as it passes over the cylinder. Although some sheet-fed machines are in use for special purposes, the majority of gravure printing is done on rotary machines. Because of the high cost of preparing the printing cylinder the process is generally only economic for very long runs, and is particularly suitable for the production of large quantities of advertising material, or for large-circulation periodicals. A succession of small developments have raised the efficiency of gravure printing machines, but these relatively small changes have occurred at a time of growing demand for coloured printing, so that the effect on employment has only been to reduce slightly the growth of employment in gravure plants. In view of this, we do not discuss technological change in the gravure process further, limiting ourselves to changes in the letterpress and offset processes.

(ii) *New Letterpress Equipment*
Although the technical advances in letterpress technology have been small compared to those of offset, investment in reel-fed rotary and sheet-fed rotary machines has of course been taking place, particularly where the demand was growing for products not suitable for production by other techniques. For example, in one of our cases new reel-fed rotaries were introduced by an organization printing mass-circulation daily and Sunday newspapers. The circulation of both had been increasing for a number of years, and the number of pages in each issue had also risen. For some time it had been necessary to begin printing earlier in the evening in order to meet delivery schedules, but as this meant some loss of topicality in the newspaper, investment in new plant was decided upon. Although the existing machinery had originally run faster, by this time it was only capable of operating at about 28,000 copies an hour; the new machines printed at 40,000 copies an hour, and were said to be capable of running at up to 50,000. A very similar increase in output between the old plant and the new was also claimed by another firm which has been modernizing its equipment for rather similar reasons.

56

But at neither of the plants studied in detail was there much *direct* change in the number of people employed in the machine room following the change. Thus at the last plant mentioned the staff on each of the new machines was 12½ men, compared to 8 on the old machines, giving a very similar number of men per unit of output. At two other plants, there was no change in the number employed when new printing machines were installed, with the existing number of men being employed for a given size of newspaper.

Apart from possible trade union opposition to manning reductions,[1] this may have been partly due to the fact that with the competitive pressures to which some newspapers are subject, great importance is placed on topicality. With fixed distribution schedules, this can only be improved by starting to print later in the evening. It follows that if new equipment of higher capacity is installed, this may not be used to reduce the number of presses (and therefore workers) required to produce a given output in the previous time period, but to reduce the printing time. For example, at one plant the new presses were used to reduce the printing time on large issues by about an hour. A similar number of machines was employed, but for a shorter period of time.[2] However, the shorter printing time would have reduced the workers' overtime earnings. Consequently, when the company announced the changed schedules, it proposed that the loss in earnings could be minimized or offset if the union agreed to reduce staffing in return for a share of the savings. Its proposals are shown in Table 3.1 which indicates a considerable reduction in the number employed on the larger issues. This reduction is, in fact, more apparent than real, due to the phenomenon known as 'ghosting'. Because of the restriction of entry to the trade, it was rarely possible to fill all of these jobs, in which case the pay due to the absent workers was shared amongst those present.[3] After lengthy negotiations with the trade union, agreement was reached on the basis of the company proposals, and substantial cuts in manning were achieved. The remaining employees benefited from a half-share of the savings, so that their earnings did not fall with the loss of overtime. Thus, although it had

[1] This is a central and recurring problem in any discussion of technological change in printing, and we devote a large part of the next chapter to an examination of it.
[2] This seems to be a fairly common phenomenon in the industry; cf. Royal Commission, op. cit., *Oral II*, 'Evidence of the Typographical Association', Q.11415–11420.
[3] Cf. Royal Commission, op. cit., *Report*, para. 97.

TABLE 3.1

*Change in the Number of Machine Assistants Required for
Various Sizes of Newspaper in One Company*

Number of Men	Number of Pages					
	20	24	28	32	36	40
Daily Paper:						
Old staffing	598	598	771	771	—	—
New staffing	535	535	535	535	—	—
Sunday Paper:						
Old staffing	—	—	683	683	834	834
New staffing	—	—	644	644	644	644

Source: Information supplied by the company.

not been possible *directly* to reduce manning on the new equipment, a reduction was nonetheless achieved indirectly.

Although sheet-fed rotary letterpress machines have not been very widely installed, two cases were examined which indicate an interesting contrast between this process and the traditional flat-bed process. From Table 3.2 it can be seen that sheet-fed rotaries save on both

TABLE 3.2

*Comparison of Capital Cost, Output and Labour Requirements of Flat-bed
and Sheet-fed Rotary Letterpress Machines* 1967

Comparative features	Process	
	Flat-bed	Sheet-fed rotary
Capital cost (£)	20,000	25,000
Labour[1]	1 minder and 1 assistant	1 minder and 1 assistant
Output per hour (copies)	2,000	6,000

Note: In Scotland only a minder has to be employed on each machine, drawing assistants from a common pool.

Source: Information supplied by companies.

capital and labour costs per unit of output. Both types of machine use the same amount of labour, though the sheet-fed rotaries have an output approximately three times greater than the older machine. At neither of the plants had there been any significant adjustment problem. One plant was a large general printing works with a great many flat-bed machines, which had replaced six of these machines by two sheet-fed rotaries. Thus four workers then did the work of

58

twelve previously, but the eight who were displaced were readily transferred to vacant jobs in the department. Those selected to work the new machines were retrained without any difficulty in less than a week. At the second plant studied, output of a certain type of work had been growing. As it was considered to be most economically produced by sheet-fed rotaries, three of these machines had been introduced in recent years. These machines were additional to existing equipment, and tended to be manned by the better operators amongst those previously on flat-bed machines. Again there were no retraining difficulties.

The introduction of new rotary letterpress techniques in the newspapers studied did not therefore cause much direct displacement of labour, with the extra capacity being used to reduce printing time and improve the topicality of the paper. However, its introduction has sometimes been accompanied by attempts to revise existing working arrangements so as to improve efficiency and these have produced considerable labour savings. More direct and obvious reductions in labour requirements result from the installation of sheet-fed rotaries, but there were neither adjustment nor training problems at the plants studied.

(iii) *Web-offset*

Although there have been few problems where new letterpress equipment is concerned, one might expect rather more difficulty with new web-offset installations, which represent a major change in the basic process. The sheet-fed versions of the offset-litho process are well established in the printing industry for relatively short runs of coloured material, and the quality of work done has benefited greatly from the improvements in ink and other material mentioned earlier. It has been estimated that there are some 7,000 such machines in operation. But much greater significance is attached to the reel-fed version known as web-offset which became available in the late 1950s and which is now an economic process for many types of output, both in the newspaper and general printing parts of the industry.

The initial stimulus to the introduction of the process has normally been a growth in demand for the product. Although the demand for newspapers as a whole is fairly static, some newspapers, particularly local ones in areas of expanding population, have experienced a substantial growth in demand from both advertisers and readers, which has outstripped the capacity of their existing plant. Similarly, demand for general printing has been expanding for some years at an average rate of 5 per cent each year, so that many firms, particularly those in the most rapidly growing parts of the industry, have also needed to instal additional plant. The principal question in recent years for many such firms has been whether to instal additional or larger letterpress machinery, or to change to the web-offset process. In some cases the output structure has itself decided the issue. Web-offset is not yet economic for very long runs of more than about half a million copies, nor is it economic for very short runs of less than about 15,000 copies.[1] But within these limits web-offset has increasingly come to be regarded as an alternative to letterpress (or gravure) printing, and a question of considerable interest is why firms have decided to make the change.

It is certainly not because of cheapness, as web-offset generally involves both higher capital and operating costs than comparable new or second-hand letterpress machinery.[2] But against this, web-offset has the advantage that illustrations in both black and white or colour can be produced much more cheaply, and to a higher quality, than is possible with letterpress printing, and the general appearance of the publication can be made more attractive to readers. This is important for the circulation of newspapers and of certain types of magazines, particularly of the trade and technical variety. But it is even more important because of its appeal to advertisers, on whom both types of publication rely for a large proportion of their revenue. Principally for these reasons a number of firms have decided to instal web-offset equipment when faced with a need to invest in additional capital equipment.

Installation of web-offset in newspapers involves the total replacement of existing printing plant, but in employment terms the change is not very serious, as firms changing to the process have continued to employ a machine-room labour force of much the same size. For example, a small daily newspaper which adopted the process had previously operated its rotary letterpress machine with one minder and nine assistants, but the new plant was operated by two minders and eight assistants.[3] The existing letterpress labour force was retrained to do the work and the Newspaper Society has reported that all of the newspapers in its membership who have

[1] These are rather broad generalizations which may not apply in the case of particular firms, but they serve to indicate the levels of output for which web-offset is a realistic competitor with alternative techniques.

[2] Cf. T. A. Margerison, 'The Impact of Web-offset on Newspaper Production', *British Printer*, July 1967.

[3] There are, of course, effects on the preparatory trades such as stereotypers and process engravers, and these are discussed in Section III.

gone over to web-offset have done likewise.[1] In the general printing industry, the labour force effects of web-offset can be more considerable. The rate of output of most web-offset machines is at least ten times that of a flat-bed machine, producing upwards of 20,000 copies an hour compared with around 2,000 from the latter. But the labour requirements do not increase proportionally with the output. At one plant, for example, machines are being operated by 3 minders and 3 assistants to produce an output which if produced on 10 flat-bed machines would have required a labour force of 20 men.

But since, as we explained above, most of the machines have been installed under conditions of expanding demand, the question has usually been one of recruiting and training the extra labour rather than one of displacement. Recruitment does not appear to have been very difficult as a sufficient number of the industry's labour force appear to have welcomed the opportunity of employment in plant using a new and expanding process. At Southwark Offset for example, a new web-offset plant established as part of a wider rationalization of the printing activities of International Printers Ltd. (a subsidiary of the International Publishing Corporation), recruitment was essentially a matter of transferring labour from other company plants.

Although the quantitative aspects of labour-force adjustment have not been difficult, there is still the question of the ease or otherwise with which the labour force adapted to the new equipment; in particular, what sort of training problems arose? In fact these have not been very serious. Although the principles of offset printing are quite different from letterpress, web-offset is very similar in one respect, particularly when installed at a newspaper plant. This is that it is printing with paper in reels as distinct from sheets, and a knowledge of the operation of reel-fed machines is considered more important initially than an understanding of the 'ink and water' technology on which offset printing is based. Thus when newspapers have changed from one process to the other, employees who have been handling the reel-fed letterpress rotary are familiar with this aspect and are able to pick up the principles peculiar to offset fairly easily. One newspaper for example retrained its labour force by having them practise on the new machine for about a month (at first under the guidance of the manufacturers), while continuing

[1] Court of Inquiry into the problems caused by the introduction of web-offset machines into the printing industry, *Report* (Cmnd. 3184, HMSO: London, 1967), para. 117. This will subsequently be referred to as the Cameron Report.

to produce the newspaper on the old machine. At the end of a month production was switched quite easily to the web-offset machines. Similarly, at Southwark, where the labour was selected from other plants in the group, the company was prepared to employ people without a knowledge of offset, provided they were able to work reel-fed machines, and provided they were fairly young. The only formal element in training was that one crew spent a few weeks at the manufacturer's plant becoming familiar with the machine. Then they returned to Southwark and, with the help of two litho-technicians (members of ASLP) who were experienced web-offset printers, trained other crews as they were recruited.

Although the installation of web-offset does not appear to have led to any serious problems so far as employment or training is concerned, it has raised a number of industrial relations problems. These, however, are part of the wider question of the role of the printing trade unions in relation to technological change and this discussion is best deferred until the following chapter.

Industrial relations questions apart, one is struck by the apparent lack of major employment problems consequent upon the introduction of the new types of printing equipment we have discussed. In quantitative terms the problem has certainly been mitigated by the continuing growth of demand for printed material, so that much investment in new equipment has been to meet a growth in demand or to improve quality, rather than to save labour. Secondly, there have been few retraining problems. Again this may partly be a reflection of the institutional barriers which rule out the use of non-craft labour on craft jobs, so that a source of labour for new equipment which may have required considerable further training is in fact not available. But a more important reason seems quite simply to be that the new equipment (which also often embodies a number of automatic controls) is generally not so different from other types of printing machine as to place it beyond the ability of a reasonably intelligent printing craftsman after a few days or at most weeks of guidance. The apparent ease with which the labour force has adapted to the operation of the new printing techniques, particularly web-offset, is perhaps the most interesting point to emerge so far.

III. THE PREPARATORY STAGES

Before printing can take place it is necessary to prepare a printing surface. We can distinguish three separate operations in these preparatory stages. The first is to set the original text in type, which may be of either lead or film. The former is the traditional medium, and the work involved is generally known as hot-metal setting. The more recently developed technique of setting type on film is known as filmsetting.[1] Whether type is set in lead or film, a computer may be used to increase the speed of the operation. The second preparatory operation is to convert any illustrative material such as photographs or diagrams into a form suitable for printing. Finally, both the text and illustrative material must be brought together and arranged into the required page layout, and a surface produced which can be put onto the printing machine.[2]

(i) *Typesetting*

Originally a compositor set type by selecting the appropriate typeface manually from a case in front of him. But since the 1890s machines have been available on which the compositor operates a keyboard, causing the required type to be formed in lead. The main machines are the Monotype, Linotype and Intertype. The former produces a paper tape with holes punched in it, each representing a character. This is then put on a casting machine which forms each character individually. Being a very flexible machine, it is the basic equipment in general printing. By contrast the Linotype or Intertype machines cast complete lines of type directly into metal and therefore operate more quickly, but they lack the flexibility of the Monotype and cannot do the same range of work. The advantage of speed has, however, made them the basic typesetting equipment of the newspaper industry.

The second method of setting lead type is known as teletypesetting (TTS) which is an adaptation of conventional Linotype or Intertype machines. Here the compositor operates a special keyboard quite separate from the Linotype or Intertype line-casting machine, producing a paper tape with holes representing characters and justification instructions (i.e. instructions on spacing and hyphenation). The tape is then attached to the line-casting machine and activates it in the same way as a compositor would if operating it directly. Output per keyboard can be substantially raised by the use of this method because less effort is required to operate the simple tape keyboard, and the line-casting machine fitted with the punched tape operates more rapidly than when an operator is working it directly.

[1] An alternative designation is 'photo-composing', but we shall use the term filmsetting throughout the discussion.
[2] The reader may find it helpful to refer back to Diagram 3.1 on page 54.

An extension of this principle enables type to be set simultaneously in more than one centre. Without this equipment it is necessary for a newspaper which wants to print identical material in separate plants to employ a man to set the type in each plant and a press telegraphist to transmit the material by wire between the plants. Using TTS the operation of a single tape keyboard linked to transmitting equipment can activate Linotype machines in the other plants without any further intervention.

The successful development of filmsetting dates from about 1956. The principle of the technique is that a compositor operates a keyboard to produce a punched paper tape. The information on the tape causes a separate photo-unit to select the appropriate characters from a matrix, and to project them in the correct sequence onto a roll of film. Several different filmsetting systems are produced, but the only distinction we need draw at the moment is between the Monophoto system and the rest. The Monophoto system is an adaptation of the Monotype hot-metal typesetting equipment and the keyboard at which the operator works is very similar in both machines. Other systems, such as the Lumitype, were designed from the start as filmsetting equipment and are not derived from earlier hot-metal methods.

One way of increasing the rate of output of keyboard operators, whether type is being set in lead or film, is to use a computer to assist typesetting. It has been estimated that about a quarter of an operator's time is spent in justification work.[1] If a computer programmed to make these decisions is installed between the keyboarding operation and the unit producing the metal or film type, this time can be saved. In more sophisticated applications it will arrange the layout of pages and the position of illustrations, and a computer is particularly valuable when complex setting and revision is involved in the production of, for example, directories and reference works.

The principles of hot-metal setting are still the basis of the typesetting operations of most printing firms, but both filmsetting and the more modern refinements of hot-metal setting have widened the range of methods which are available. TTS has been introduced by a number of firms where the typesetting load at a plant has for some reason increased. This can be met by either installing additional conventional line-casting machines and employing more compositors, or by installing equipment for TTS setting. A new line-caster costs about £8,500 and the TTS keyboards and adaptations about £2,000.

[1] See J. E. Reeve Fowkes, 'Productivity in the Composing Room', *British Printer*, April 1968.

The TTS equipment is not subject to rapid obsolescence, so the annual depreciation charge can be quite low and the labour savings obtainable have often made its installation economically attractive.

The installation of TTS raises the output of the keyboard operator. In one of our cases the increase in speed was fairly modest, from about two lines per minute to two and a half lines, but this can be at least partly explained by rather peculiar working arrangements in that plant.[1] Another plant, which is more typical, found that the output of operators on tape keyboards was double their output on conventional line-casting machines. This increase in keyboard productivity is partly offset by the need to employ men to supervise the operation of the tape-operated line-caster. When the tape has been produced by the keyboard operator it is then put on the line-caster. The output of a tape-operated machine is from eight to twelve lines a minute and, therefore, takes the output of two keyboard operators. The normal practice in England is that one man supervises two tape-operated line-casters, but in Scotland until recently the agreement with the trade union made it necessary to employ one man on each machine. Consequently, in the examples quoted above both of which had four tape-keyboards and therefore two tape-operated line-casters, two men were employed to supervise the latter machines. By comparison with practice elsewhere this was clearly excessive, but following the implementation of a new agreement in 1968 which allows one man to operate two machines it has been possible to reduce this number, so increasing the benefit from the machines.

The main adjustment problem is that of retraining operators to work efficiently on the rather different, though simpler, keyboard. At the second firm mentioned above the operators went on a typing course for six weeks, then built up their speed during production work in the plant. At the other plant, the peculiar arrangement referred to above is that every man in the composing department takes a turn on the TTS keyboards for a fortnight at a time. This is a house agreement negotiated with the union concerned and must clearly have contributed to the low increase in output per man which that firm has achieved on the equipment. Wages are slightly higher on this equipment than on conventional Linotypes or Intertypes, being about 30s higher at one plant and 20s at the other above the basic rate for the machines.

TTS equipment can also be used to set type simultaneously in more than one centre. This application is obviously rather limited and

[1] See next paragraph.

most firms who could use it already do so. One installation occurred when *The Guardian* began to print in London as well as in Manchester; to save the time and expense involved in setting material twice it linked the two plants by a TTS system.[1] Because of the reduction in labour from two compositors to one telegraphist a dispute arose between the composing unions who set the type and the press telegraphists who transmitted it, over which union's members should operate the keyboards. After three years a formula was eventually put into operation which allowed the manning of the equipment to be shared by members of the two unions, which have now merged.[2] The installation enabled an increased volume of work to be handled with the same staff.

The introduction of *filmsetting* has been relatively slow. Although the equipment has been available for a decade, and although there are some 6,000 firms in the industry, most of which have typesetting departments, an inquiry by the British Federation of Master Printers at the end of 1965 revealed only 51 companies with filmsetting installations. Even where it has been introduced it is not necessarily the only technique, as a firm may continue to use traditional hot-metal equipment with film-setting as a supplement. The use of the technique is by no means limited to large firms and many very small establishments have in fact made use of it, although a common feature of those who have installed it is that they have some offset printing equipment.[3]

There are, however, several advantages in using film. To make an offset plate it is necessary to produce a film of the images to be printed and it is obviously more convenient to set type straight on to film. The quality of type is superior, being clearer and having a better definition. Furthermore, setting which includes graphs or complicated type arrangements is best done on film as the make-up of a page which would be almost impossible using lead type can be done very easily with pieces of film. These two reasons weighed very heavily, for example, with one typesetting trade-house, which by installing filmsetting equipment was able to meet demands from some printers for better quality type and more complex layouts. Another firm which was gradually changing its printing method from letterpress to offset also installed some filmsetting equipment, to meet

[1] Cf. Royal Commission, op. cit., *Documentary IV*, Evidence of the National Union of Press Telegraphists, para. 16.
[2] Cf. Royal Commission, op. cit., *Oral II*, Evidence of the National Union of Press Telegraphists, Q.9211–9219.
[3] Based on information supplied by the BFMP together with information about the firms concerned obtained from trade directories, etc.

the increased demand for typesetting, alongside existing hot-metal equipment.

Newly established offset printing plants in particular, with no existing hot-metal equipment, are likely to set on film from the outset. For example, we mentioned earlier the establishment of Southwark Offset in London in 1964. This was to be a major, new web-offset plant at which an attempt was being made by the company to break away from many of the industry's traditional practices. As well as being technically suitable, filmsetting fitted naturally into this concept and hot-metal setting was probably never seriously considered. Another case study was at a new web-offset factory established outside London. For some years the type for this new plant had been set in hot metal at another factory in the same group, but as the web-offset plant grew in size it eventually justified its own setting department. It was decided to set by film for two reasons. The first was that it was considered the more suitable process in a new offset printing plant without previous setting equipment. The second, rather interesting, reason was that being a relatively new process there was more room for subsequent improvement. The balance of advantage in terms of cost was expected to move in favour of filmsetting. Being a completely new typesetting operation, it was felt that it was better to start by using this process, as it would be easier to change to a later model from an existing filmsetting installation than from hot metal.

Filmsetting does not alter the fundamentally manual nature of the typesetting operation. Although more advanced scientific principles are used to produce type in the machine than in hot-metal equipment, it still depends on an operator depressing the appropriate keys on a keyboard. Firms which have installed the Monophoto system, an adaptation of Monotype hot-metal equipment, find that output per man is practically the same on the two machines. Other systems are, however, based on a typewriter keyboard. With teletypesetting, a compositor's output can be doubled if he operates the simple typewriter keyboard rather than the complicated hot-metal equipment, but when the same type of keyboard is used for filmsetting, an equivalent improvement is not achieved. The reason for this is that the cost of corrections is greater with film than with lead type, and therefore greater care must be taken during setting. It is also necessary to employ labour to operate the processing units which actually produce the film-set type. These process the tape from several keyboards (the number depending on the model used) and it is normal to employ one man to every two processing units. The

67

result is that filmsetting does not normally have any very significant effect on the number of men required to produce a given output —its advantages are those of quality and convenience, rather than of saving labour.

When firms have installed filmsetting they have normally retrained existing compositors. Where Monophoto machines have been installed in place of Monotype equipment, the adjustment is very easily accomplished as the two keyboards are very similar. Firms naturally tend to select their better operators to use the new equipment, and the transition is usually made in a few days. The adjustment has been a little more difficult where firms have installed systems using type-writer-based keyboards. These are quite different from the hot-metal keyboards with which the operators were previously working. The normal experience seems to be that it takes two or three months to achieve a satisfactory rate of output while the operator becomes familiar with the new keyboard. One firm which ran into particular difficulties solved the problem by taking the men off production work and putting them through a special course of training. For several weeks an instructor tried to raise their speed by putting them through a typing course on ordinary electric typewriters, before putting them back for a further period on to the proper keyboards. This appears to have raised output in a matter of weeks by about 40 per cent, quickly repaying the cost of retraining.

This raises a question which has been discussed in the industry of whether it is really necessary to employ retrained compositors for this work. It has been forcibly argued[1] that since the basis of most types of filmsetting equipment is a keyboard based on conventional electric typewriters, the basis of an operator's training should be the same as for an ordinary typist. When competence in this respect had been achieved, the 'typographic' elements of the job could then be added. The conventional view in the industry, particularly of the trade union most concerned, stresses the typographic elements of the job such as the knowledge of type faces, layout, etc., arguing that only skilled craftsmen possess these skills, and successfully opposes the use of other types of labour for this work.

Finally, *computerized techniques of typesetting* have not been widely introduced in the printing industry, and we did not study their effects on employment in any detail. In general, however, they have the effect of reducing the amount of composing labour required, to

[1] As in, for example, 'An Analytical Approach to Typesetting', *British Printer*, February 1967.

an extent ranging from about 25 per cent when they are used only for making justification and hyphenation decisions, to very much greater proportion in the more advanced applications. If computers were to be used in typesetting, there would, of course, be a growth in the number of programmers and other ancillary computer occupations, partly offsetting the decline in traditional composing labour. Computer setting may also erode the virtual monopoly which craft compositors still retain over typesetting operations. To the extent that original copy has simply to be typed on to a magnetic tape which then, in accordance with previous instructions as to type-face and page-size, prints out a tape ready for type-casting or film-processing machine, it does not seem that conventional composing labour is really required. In this form typesetting is very similar to many other commercial applications of a computer in which information is generated and fed into the computer by relatively unskilled labour.

Although one would expect computers to take over fairly quickly much of the listing type of work for which they are particularly suited, and even to stimulate its growth, this type of work constitutes only a small proportion of the output of the industry. Over the greater part of the industry there is little sign yet of a rapid acceptance of the technique. In part, this may be the result of caution in the face of so radical a development, and unwillingness to be among the first to use a technique which will certainly have many initial difficulties. But in addition there may be uncertainty over the economics of the process. Not only is the initial cost high, but the basic and peripheral equipment is still evolving rapidly so that obsolescence promises to be rapid. Consequently, depreciation charges will be heavy. In addition, for many types of output the savings or benefits from computers may be considered inadequate—typesetting labour is reduced, but in cost terms this may not make the investment economically worth while.

In summary, as with most new installations of printing machinery, the principal motive for the introduction of new typesetting equipment has been a growth in demand. Particularly where TTS equipment with quite clear labour-saving effects has been installed, this growth in demand has prevented any serious problems of redundancy; the increased labour productivity has served to prevent an increase in the work force, rather than to reduce it. Filmsetting does not generally save any significant amount of labour, as the essentially manual keyboarding operation remains, though systems based on typewriter keyboards do appear to have a higher output per man than conventional keyboards. The adjustment to the newer type of keyboard has caused some difficulties, particularly with older workers, but a fairly short period of retraining on, for example, a typing course appears to have been all that has been necessary.

(ii) *Illustration Work*

We have so far considered the several alternative ways of setting type, and we now turn to the preparation of illustration material. As Diagram 3.1 showed, the work required to transform the original illustration into a printing surface depends upon the printing method used. The most important technological changes in this field have been those affecting the process engraving operations related to letterpress printing. The traditional way of making process-engraved blocks for letterpress printing has been supplemented by what is known as the powderless etching technique. The essential distinction necessary for our present discussion is that unlike the old technique which requires considerable manual attention to the work during engraving, the new process can continue without attention until etching is completed, and requires considerably less labour. The equipment for this process only costs about £1,200 to install, and the very substantial labour savings which are possible have ensured its very rapid adoption by newspapers, general printers and trade houses. For example one writer has pointed out that while in 1955 only one company in the United Kingdom was using powderless etching equipment, by 1964 this had risen to over 200, with many of these using more than one machine.[1]

Estimates vary as to the reduction in etching time which is achieved by using powderless etchers. In evidence to the Royal Commission on the Press, a trade union official spoke of a reduction from $2\frac{1}{2}$ hours to 15 minutes in the time required,[2] while another estimate[3] is that a line block could be etched in 7 minutes compared with 45 minutes by the old method. Clearly there is a substantial reduction in the labour content of an engraved block[4]. Not

[1] Eric A. Williams, 'A Decade of Powderless Etching', *Penrose Annual*, 1965.
[2] Royal Commission, op. cit., *Oral II*, Q.9726.
[3] Cf. C. M. Flint, 'The A.N.P.A. modification of the Dow Process', *Penrose Annual*, 1955.
[4] The reduction in etching time may be slightly offset by the greater care which is required at the negative stage. Using the old method the etcher worked on the block for a considerable time, and was able to influence materially the quality of the block. But with the powderless etchers the operators' control is greatly reduced and corrections have to be made at the negative stage.

70

only does powderless etching reduce the time taken to produce a single block: it also makes it possible to etch several simultaneously. Previously each plate was etched individually, but with a large powderless etcher, a firm can wait until it has enough individual blocks to fill the machine and do them all together, reducing the total etching time even more.

The consequence of this is that unless the output of engraved blocks is increased fewer process engravers are required. Some newspapers have used the shorter time taken to produce a block to increase the number of pictures they make. For example, it is possible to change pictures more frequently between editions than was possible with the older, slower technique and this has absorbed some of the new equipment's capacity. But the scope for this is limited. One newspaper, which has ten process engravers, has installed the equipment and the management considers that it could easily produce the output with seven or at most eight operators. But it has been prevented from making the adjustment under the terms of a 'no-redundancy' agreement which it has made with SLADE. Furthermore the union has insisted that the firm replaces men who leave. One man left during 1966 and the firm naturally did not want to replace him. The union, however, insisted that a replacement should be found and eventually the firm accepted this and took on another man, so that it has not been possible to adjust the labour force to the requirements of the new process.

In general printing the demand for process-engraved blocks has been fairly static. These are, of course, only required for letterpress printing and the growth of web-offset printing has therefore reduced the demand. The capacity of the industry has been raised substantially by the use of powderless etchers, and faced with a static demand, many firms have gone out of business. The London branch alone of SLADE has calculated that about 400 of its members (about 15 per cent of the branch strength) were made redundant in the period 1961-6, as a result of closures or mergers of process-engraving establishments.

The labour force consists of men with a highly specialized craft training for whom a decline in employment opportunities could have serious consequences. Conscious of this, their union has actively tried to maintain the number of jobs available to them. In 1960 an agreement[1] was reached with the employers that two men should be employed on each machine in the general trade, although no such agreement was ever made for the newspaper industry.

[1] Process Trade Agreement, 1960.

71

It was recognized that these machines were generally operated by one man on the Continent,[1] and this agreement was subsequently replaced in 1965 by one which allowed single manning of the equipment.[2] The union has also put pressure on employers not to declare redundancies, or at least to delay reduction in staff until the union could place men elsewhere. In this way it has undoubtedly slowed down the reduction in etching staff which would otherwise have occurred, although numbers have still fallen.[3]

Faced with this decline in jobs for an important sector of their membership, the union has adopted two policies to try to overcome the problem. First, entry to the sections affected by technological change has been limited. The 1960 Process Trade Agreement drew a distinction between the various trades an apprentice could be engaged in, and for those most affected by technological change the quota of apprentices was fixed at one for every eight craftsmen, while for the rest the quota remained at one for every five craftsmen. The 1965 Agreement subsequently extended the smaller quota to all occupations though it was intended to revise it to one to seven at a later date. However, the acute problems in the London Branch to which we referred above resulted in a three-month suspension of all intake to the vulnerable line and half-tone etching occupations in the London area.[4] The union tried to have this suspension continued for an indefinite period, but were unsuccessful in this.[5] Secondly, there has been a steady transfer of men from the process-engraving trades to those trades which the union controls and which are expanding, such as gravure cylinder making.[6] These policies have the effect of reducing the supply of labour in the vulnerable trades, and together with efforts to maintain demand have prevented any significant unemployment from arising amongst photo-engravers, although their numbers has fallen and will continue to fall.

(iii) Preparation of the Printing Surface

When type has been set and illustrative material prepared, it must then be arranged into the required lay-out, an operation known as page make-up. If the type has been set in lead and is to be printed by a letterpress process the page make-up work is followed by the

[1] Royal Commission, op. cit., *Oral II*, Q.9732.
[2] Process Trade Agreement, 1965.
[3] See Chapter 4 below, p. 91, Table 4.1.
[4] SLADE *Annual Report*, 1965, p. 6.
[5] SLADE *Annual Report*, 1966, pp. 6 and 7.
[6] Ibid., pp. 12 and 13.

production of a stereotype. This is a duplicating printing plate of harder metal or plastic necessary to avoid the deterioration which would occur if the soft lead type were used on the printing machine, and of course to enable several printing machines to print the same material simultaneously if necessary. If type set in hot metal is to be used for offset or indeed gravure printing, it is necessary to convert the type into film form which can then be used to produce the printing surfaces.

Technological changes have not greatly affected the operations necessary to produce the actual printing surface, although new materials have become available which simplify the production of letterpress printing plates and reduce the time taken to fit them to the machine. Furthermore, the production of offset printing plates has been greatly simplified by developments in materials and in the equipment used. Some of the modern equipment is being used by people trained in other crafts who were very quickly retrained in this new occupation. Apart from this, developments in plate-making have been of a relatively small and continuous nature. But two problems, one a direct and the other an indirect effect of technological change, have affected those involved in the preparation of the printing surface.

The first is a demarcation problem. The traditional arrangements for making up pages have been that the whole of the make-up of type and illustrations for letterpress printing has been done by the compositors (members of NGA), while in offset and gravure they have only been concerned with the lead type; as soon as this has been converted into film it has been taken over by SLADE members who have arranged both it and the illustrations into page form. A problem arises when, for example, a letterpress printer changes to offset printing and installs filmsetting. The NGA members try to follow the work irrespective of the process used, but this brings them into conflict with SLADE, who in offset printing do much of the make-up work. A demarcation agreement was worked out in 1958, which allocates the make-up of film-set type to the NGA, while the positioning of illustrations is done by members of SLADE. Although this seems a very cumbersome arrangement, firms seem to find it more of a minor nuisance than a real hindrance to efficiency. In any case it does not seem to have been widely observed, for when SLADE made a survey[1] of current manning arrangements in filmsetting installations, only 12 of the 50 plants examined were manned in complete agreement with the 1958 formula. Fifteen of the others had

[1] SLADE *Annual Report*, 1965, p. 18.

73

some special circumstance, but at the other twenty-three the Society was concerned at the departure from the agreement, in that NGA members were doing work properly due to the SLADE members.

It appears that there has been a reluctance on the part of employers to follow the formula rigidly, and 'there has been a natural tendency for both employers and employees to rationalize production methods, sometimes at the expense of traditional demarcation boundaries'.[1] As a result of pressure from the NGA, a new interpretation of the 1958 formula was agreed in May 1966 which in effect allows for a more flexible interpretation of the agreement, but allowing NGA members to do more work than they were entitled to under the previous agreement.

A related question is whether the make-up of filmset type is most efficiently done by men who have been used to handling lead type. Several companies have argued that the work of sticking pieces of film on to a page layout is a job requiring considerable manual dexterity and could probably be done better by women. Again the NGA and STA have resisted this, and all the companies interviewed in this study employ men for the work. But two firms who operate printing sections as a service department to their main business in engineering are understood to employ women satisfactorily on this work. Further evidence that trained compositors are not really necessary for this work is suggested by the experience of Southwark Offset. Of the first 39 men employed on the work only 16 were previously compositors, although another 8 were keyboard operators who may have done some make-up work. Of the others, 9 were previously machine minders, 4 were process engravers and 2 were in other occupations.

The second problem arises from the close link which exists between the preparatory operations and the printing method. Even if the technology of these operations is relatively static, employment may be affected by a decline in the use of the printing method to which they are an ancillary process. For example, the stereotype used for letterpress printing is not required for offset work where the surface is produced by printing the made-up film on to a pre-coated plate. The latter operation was once a skilled job but now appears to have been made much simpler, and there are several cases where a firm has retrained its stereotypers to make offset plates. The retraining appears to have presented no problem, being achieved in a matter of hours.

An identical problem has further reduced the employment

1 SLADE *Annual Report*, 1965, p. 18.

opportunities available for process engravers, who, as we noted above, have experienced a substantial loss of jobs as a result of technological changes in process engraving. The method of type-setting does not necessarily have to change when offset replaces some other printing form; existing hot-metal equipment can continue to be used if suitable adaptations are made. But the production of illustrative material is entirely different, requiring simply a negative of the original, which is used to make the printing surface. The process-engraved blocks used in letterpress work are not required. To the extent, therefore, that offset printing takes over from letter-press, the process engravers suffer a further reduction in demand for their services. The very close relationship between the printing method and the production of the printing surface means that in considering the labour-force requirements in this part of the pre-paratory stage, we must take into account not only technological changes within these processes, but also the effect upon them of changes in the printing method.

IV. THE FINISHING STAGE

After printing, the product has to be finished and packed in readiness for dispatch. In contrast with earlier stages of production the finishing operations have traditionally involved a considerable amount of manual work and little capital equipment. About one-quarter of the industry's manual workers are employed in these operations, the majority being non-craft workers. The range and type of work to be done is as diverse as the products of the industry. Nonetheless, the effects of some of the technological changes which have recently been introduced into this work can be seen by looking at a number of particular cases in the newspaper and general-printing parts of the industry in turn.

(i) *Mechanization of Newspaper Publishing Rooms*
Newspapers leave the printing machine in a completely finished form, but must then be made up into bundles, wrapped, tied and conveyed to a delivery van. These operations take place in the 'publishing room' of a newspaper, which in effect lies between the end of the printing process and the first stage of the distribution system. The rate of output of the publishing room is closely geared to the output of the printing machines. If these machines are able to start printing later in the evening as a result of higher output rates, the rate of output of the publishing room must also be

raised to pass the newspapers more quickly into the distribution system.

The volume of material to be printed and packed varies considerably from day to day. Although the number of newspapers produced in a plant each day remains fairly constant, the number of pages in each issue fluctuates widely. The number of pages is largely determined by the amount of advertising, and this tends to be concentrated on certain issues, which in consequence are much larger than those with less advertising. But in order to keep the printing machines running at their normal speed and to fit in with the very tight time limits of the distribution system, the greater volume of work must be handled in roughly the same time as the lesser volume. The practice has therefore developed of varying the number employed in the publishing room in accordance with the size of the newspaper. For example, at one plant 545 men were employed if the paper contained 24 pages and 688 if there were 32 pages. To meet these fluctuations the permanent staff of regular employees was, and to a lesser extent still is, supplemented as required from a pool of casual labour.

This pool for national newspapers is operated by the London Central Branch of SOGAT, which has been able to establish a strong control over the supply of labour to both general printing and newspaper firms. There is virtually no non-union labour and union members wishing to work in central London have to join the branch. By the refusal of admission to new applicants, the supply of labour can be readily controlled. Members of the Branch working in newspapers are of two broad types. There are the employees who are in regular and continuous employment with one firm, but to meet the peak loads described in the previous paragraph the union also supplies labour on a nightly casual basis. The pool is made up either of workers for whom this is their sole occupation and who in the course of the week may work for several different newspapers, or of men who are already in full-time employment, for example with a general printer, but who may work one or two nights a week on a casual basis for a newspaper. The procedure is that when a newspaper has decided the number of pages it will print, and therefore knows the number of men in excess of its regular workers that it requires, it passes this information to the branch. Provided labour is available the branch then allocates the requisite amount to the firm.

Almost all of the national newspapers have installed or are installing mechanized publishing equipment,[1] often as part of a wider

[1] The Economist Intelligence Unit, op. cit., p. 198.

reorganization of printing facilities. In two cases the companies were concentrating printing facilities into a new plant and closing down separate units. This gave a much greater output from the presses and it was decided to install mechanized publishing equipment in the new plant. Another company installed it primarily to handle a growing peak load on one night of the week. This peak was continuing to rise and additional resources were necessary to handle the papers in the publishing room. Additional labour might have been used but the physical characteristics of the building prevented this and it was therefore decided to mechanize the operation. Other firms have installed it simply to handle the same output more economically than previous hand-tying methods, as the savings in manpower which can be secured by the use of these machines, provided output is above a certain level, are sufficient to justify the capital cost involved.

Precise measurement of the manpower savings possible with mechanized tying is difficult because the revision of manning scales for the new equipment takes place in an industrial relations environment which allows for considerable discretion and bargaining in the establishment of the number of men required. Consequently, the reductions which have been observed after the introduction of mechanized tying equipment will be the outcome of bargaining as well as of changed technology.

Mechanized tying equipment had been installed in 1961 by one plant, but staff adjustments were governed by an agreement which allowed reduction only by a very slow process of natural wastage. Consequently it was employing a staff substantially in excess of its requirements.[1] In 1963, however, a new agreement[2] allowed it to remove this surplus and to employ uniform staffings on each size of paper instead of a fluctuating one. Before this the plant had employed 337 people on large sized issues in the publishing room, but after the removal of surplus staff and some additional revisions it employed only 221. During the course of the negotiations on the changes at this plant, pressure was put on the company to make similar increased payments to comparable workers at another plant which was still using hand-tying methods. This stimulated the company to install mechanized tying equipment at the second

[1] To prevent the surplus labour becoming integrated with the labour force actually required and perhaps being regarded as part of it, the company adopted the practice of sending them home as soon as they had reported for work although subsequently paying them as if they had in fact worked.
[2] These argeements are discussed more fully in Chapter 4, Section IV, below.

77

plant as well, reducing the labour requirements there on any one night from 312 to 142.

Another installation was at a plant used to produce two papers, one on Sunday and the other on week-days, the capacity being determined by the larger Sunday paper. As the latter's circulation rose, the output of the presses began to exceed the ability of the publishing room to handle the load. As the available floor area was restricted in size it was not possible to increase the labour force significantly, and the immediate solution was that printed papers piled up until the men could handle them. This was clearly unsatisfactory and two mechanized tying lines were therefore introduced to coincide with an anticipated increase in output in September 1964, and a further two have since been added. In August 1964, a labour force of 294 was necessary to deal with the then current output of 750,000 copies of the 40-page paper. Regular employees numbered 254 at the time, the balance being made up by casuals. When the first two tying lines were introduced, the labour force on Saturdays rose from 294 to 306 but output had risen to over 900,000, representing a reduction of about 15 per cent in the labour force per unit of output. When four tying lines were introduced the labour force came down to 278.

A final illustration of the effect of mechanized tying equipment on publishing room labour may be quoted. To pack a 24-page issue of another national newspaper previously required 325 men. This was reduced by negotiations in 1964 to 280. With the installation of mechanized tying lines the number was reduced to 168. This includes 78 ancillary workers, the same as previously, which means that the number engaged on the actual packing operations fell from 202 to 90. From these examples it is apparent that mechanized tying equipment in newspapers enables substantial savings of labour to be made. The operation of the new equipment is very simple and the men became familiar with it in a matter of hours. Each plant, however, has had to increase its maintenance staff to look after the equipment. For example, the first company discussed has increased its employment of engineers by 26, and of electricians by 15, to maintain the new equipment.

Once a firm had secured an agreement with the chapel or with the branch on the post-mechanization labour force, the adjustment process was very simple, as it merely ceased to call in casuals from the union branch. The branch had taken several steps to forestall serious unemployment amongst its members as a result of these changes. It had now accepted technological changes as a means of increasing its members' earnings, but could only ensure this by strict control over the number available for work. It was, therefore, trying to reduce the membership of the branch in line with the decline in the number of jobs available in central London. It was able to prevent any new members from entering the branch and was encouraging members already in to leave the area and go to jobs outside of London. Further, it tried to eliminate the practice which had grown up whereby many of its members worked a sixth night in addition to their normal week's work. This would effectively increase the number of jobs available. It hoped to supply firms with their increased requirements of regular labour by persuading a sufficient number of casual workers to accept regular employment, and supported the action of one newspaper which persuaded about 20 of its older employees to retire. The firm had negotiated successive cuts in that labour force which eventually meant that numbers would fall below the number of men in regular employment. The expedient then adopted was to take advantage of the high average age of the labour force;[1] by making substantial compensation payments, a number of men over the normal retirement age were persuaded to retire.

(ii) *Mechanization of Finishing Operations in General Printing*
Although the products of the general printing industry are extremely diverse, books and magazines probably account for most of the finishing work of the industry as most other printed material leaves the machine in a virtually finished form and needs only to be packed and dispatched. But books and magazines generally require a series of further operations such as gathering together several different sections in the correct sequence, binding them together and trimming them to the correct size. Several new types of equipment are now available to the industry which have made possible the mechanization of these processes.

An immediate problem is to clear printed material away from the machines, so that printing is not slowed down, and this problem becomes more acute as faster printing machines are installed. One firm installed seven *counter-stacking machines* which take printed magazines from the gravure printing machines and stack them automatically into bundles of the required quantities. The firm received two benefits from the installation. First, the company had installed new and faster gravure printing machines but their actual output was governed by the rapidity with which men could lift

[1] See page 42 above.

material from the end of the machine, and this speed was lower than that at which the machine could print. The obvious solution of employing additional men to handle the material was precluded by the physical characteristics of the machine. The stackers automatically removed copies from the press and counted them for the first finishing operation, thus eliminating the bottleneck and allowing the machine to run at a higher speed.

The second advantage was that the investment should reduce labour costs quite substantially. The number of men previously employed on the work varied from 5 on some machines to 10 on others. The work was semi-skilled and repetitive. With the new equipment, one machine was cleared by 4 men and the rest by 6, and the total employed on the work over three shifts was reduced from 159 to 121, their functions being to ensure that the machine is operating satisfactorily and to move the counted stacks to a storage pallet awaiting the next production operation.

The company guaranteed that no one would be dismissed because of the introduction of new techniques. Labour turnover is low in the industry and the movement of employees between departments is restricted. These two considerations together with the no-redundancy policy, made it more than usually desirable for the company to plan ahead if it was not to be faced with the costs of continuing to employ a substantial surplus of labour once the equipment was installed. On the other hand, it was helped by a long period of advance notice of the installation: some eighteen months elapsed between the decision to order and the planned installation date. It therefore suspended recruitment to the section and when the opportunity arose, transferred people to vacant jobs in other departments. Together with some natural wastage, this brought the numbers down to the required level in the planned time.

The success of the run-down in fact led to difficulties. Obviously, as crews ran down and were not replaced the pressure of work on the remaining men increased in the period preceding the installation of the equipment. It was hoped, however, that the labour force would reach the required level at about the same time as the new equipment was due to come into operation. This did happen, but unfortunately technical troubles with the equipment delayed its operation. Those directly responsible for production wanted to take on temporary labour to meet the situation until the new equipment was in operation. Management felt, however, that if the labour force was increased it would be difficult to reduce it again, and successfully resisted these pressures until the equipment was eventually installed and the reduced labour force was then able to handle the load without difficulty.

As part of their agreement with the SOGAT chapel in the plant on the installation of new techniques, the firm had agreed that any resulting savings in labour costs would be divided equally between the company and the remaining employees in the department concerned. The savings in labour costs were determined by the earnings of those displaced: when divided amongst the remaining members of the finishing department (not just those on the stackers) the amount involved was approximately 4 per cent of the individual's basic day wage.

Considerable reductions in labour requirements were also achieved by equipment installed at the delivery end of the web-offset presses at a major new plant. At first 5 men were employed to take copies from the press, count them and stack them into bundles, but one machine was fitted with equipment which does this work automatically and requires only one man in attendance. The equipment was installed during the build-up of the labour force at this new plant and adjustment problems did not arise as the 4 men displaced simply moved to newly installed machines. In contrast to the equipment mentioned earlier, this machine has the disadvantage that it operates at a slower speed than the printing machine: consequently, the speed of the latter has to be lowered to that of the stacking equipment. It is anticipated that these technical problems will be overcome and that a more sophisticated machine will be installed on the other printing machines in the plant.

The plant which installed the stacking machines subsequently linked the stacking equipment to the trimming machines so that there is a continuous flow of magazines from the printing machine into the trimmer, eliminating one handling operation. Some labour is saved, and another advantage is that as the trimming machines are able to work at the same speed as the printing machines, the space previously required for storage of material awaiting trimming is now released for other purposes.

There was considerable delay before an agreement between the firm and SOGAT was reached on the operation and manning of the machines, but this was eventually secured and the equipment is in operation. The labour force on the operation has been reduced from 5 to 4, which appears to represent some overmanning of the equipment as the actual manual element has been eliminated. But despite the low savings in labour as a result of trade union pressure, the equipment is operating presumably because the disadvantage of overmanning is at least offset by the other advantages mentioned earlier, namely the greater speed at which the trimming machines can operate and the savings of space.

The effects on the labour force of linking several machines together can be observed at another plant producing clothbound books. As demand grew, particularly for some items with very long runs, the firm sought ways of increasing the efficiency of one of its two finishing lines, each of which consists of three principal operations. By linking these operations together, the work of removing material from one machine, moving it and feeding it into the next is eliminated and considerable savings in handling are achieved. The labour force on the unlinked line consists of 12 people, compared with 8 on the linked lines.

The same firm has also introduced two *mechanized perfect binding machines* for gathering, binding and trimming most of its output of paperbacked products. A book or thick magazine is usually printed on several separate sheets of paper each containing 32 or 64 pages. After printing, these are folded and gathered together in the correct sequence, bound in some way, the cover put on, and finally trimmed to the required dimensions. Each of these operations can be performed in several alternative ways requiring different inputs of labour and capital, but the main innovation of the equipment is in the actual binding. Traditionally the separate parts of the book have been sewn or stapled together, but an alternative now available is a method known as perfect binding. Here, instead of the complete sections being bound together, the back edge is trimmed and each sheet separately gummed before being placed inside the cover. Until 1959, this plant's output of paperbacks, of which the biggest item was a large-circulation monthly magazine, was gathered and stitched on a machine installed in 1935. In 1959, however, the publishers of the magazine decided that their publication would in future be bound by this new method and the printers therefore had to install the necessary equipment.

The department's peak output is reached on the eight days each month during which 1·6 million copies of the magazine are printed. For the rest of the month the equipment is used for other paperback work. With the installation of the first perfect binding machine, capacity was well in excess of demand, but increasing paperback production eventually used up the spare capacity and a second machine was installed in 1966. Capacity, therefore, again exceeds demand but it is anticipated that the volume of paperback printing will continue to grow and that the availability of capacity puts the company in a strong position to secure a share of this growth. It is worth noting that even the introduction of the second machine has not led to the displacement of the original 1935 machine; it is still in use though mainly for shorter production runs.

The output and manning of the three versions of the equipment are shown in Table 3.3. The principal reason for the considerable

TABLE 3.3

Output, Employment and Capital Cost of Successive Binding Machines

Comparative Features	Machine Purchase Date		
	1935	1960	1966
Copies per minute	150	260	350
Number employed	15[1]	13	15
Approximate capital cost	NA	£110,000	£150,000

Note: Includes 3 women employed on a separate piece of equipment, incorporated with the main machine on the two later versions.

Source: Information supplied by the company.

increase in output between 1960 and 1966 is attributed to advances in mechanical engineering which have improved the driving mechanisms and enabled the machines to withstand the stress of higher speed running. Significant developments in adhesive technology have also made a contribution. Both the later machines differ from the 1935 one in that they incorporate cutting equipment for the final trimming work, whereas the earlier machine only gathers and binds, the books then being trimmed on separate equipment. It is unlikely that the later machines could operate at the speeds shown if the cutting machine was not linked to the binder: if material had to be removed manually from the binder as in the 1935 machine this would almost certainly reduce the operating speed.

It is apparent that while output per minute has increased substantially with each successive machine, labour per unit of output has fallen. The incorporation of the cutting machine into the binding equipment on the 1960 version reduced the labour force, as the three girls taking work from the binders were no longer required. The increased output of the 1966 model compared with the 1960 one required only two additional people, one more filling the gatherer and one more porter, both reflecting the greater physical volume of work being done by the machine. The manning of the equipment in this case was not a source of serious controversy, and the number employed on the work is very close to that which the firm originally proposed.

83

LABOUR PROBLEMS OF TECHNOLOGICAL CHANGE

This concludes our examination of the case-study material on the effects of technological change. With the exception of the process engravers and perhaps those employed on packing newspapers, the technological changes so far introduced do not, in the cases examined, appear to have resulted in many serious difficulties. Because of the growth in demand, displacement has been negligible (apart from the exceptions mentioned above) and in addition few re-training problems have occurred. The only serious industrial relations issue which has disrupted production has been that over the manning of web-offset presses, but even that was a fairly isolated incident viewed against developments in the industry as a whole. But this whole issue of the trade unions and technological change is one to which we have made only passing references in the present chapter and is one of a number of wider problems to which we shall now turn in Chapter 4.

CHAPTER 4

THE ADJUSTMENT TO
TECHNOLOGICAL CHANGE

I. INTRODUCTION

The previous chapter looked at the effects of technological change on employment in the printing industry from the point of view of a necessarily limited number of case studies. But as we pointed out in that chapter, a true assessment of the labour-force effects depends on developments in the industry as a whole, and it is the purpose of this chapter to set the case-study evidence within this wider context. We begin in Section II by considering the factors which have stimulated or delayed the adoption of the available new techniques by the industry. Section III draws together the main points which have emerged in relation to the effect on employment, and in particular puts these against the aggregate changes in labour-force structure shown in the official statistics. Finally, in Section IV we consider the role which the trade unions in the printing industry have played in relation to technological change.

II. THE RATE OF ADOPTION OF NEW TECHNIQUES

Although very little research and development work is done by the printing industry,[1] it has benefited from research and development elsewhere. The established suppliers of printing equipment have produced more efficient versions of their existing machines, or have developed new equipment such as filmsetting units. Similarly, the manufacturers of paper, ink and other materials have produced better products which allow either faster or better printing to be done, again benefiting the industry. Finally, fundamental advances in scientific knowledge, particularly of electronics, have been exploited both by existing suppliers and by new ones not previously connected with the industry. For example, both filmsetting and

[1] Information made available by the Ministry of Technology shows that in 1964–1965 the paper and printing industries *together* spent only £3·1 m on research and development, less than 1 per cent of that in all manufacturing industry.

85

powderless etching equipment are the result of new scientific knowledge being applied at least in some cases by new equipment suppliers. But there have also been innovations based on the use of existing knowledge; for example, quite substantial labour savings have been achieved by the application of well-known mechanical handling principles to the finishing and packing of printed products.

However, the invention and development of a new method of production is no guarantee of widespread use. Many innovations which are technically efficient have not been widely used because they have been unable to satisfy tests of economic efficiency. In the printing industry, for example, the powderless etching technique won rapid acceptance and has been installed by many firms, while in contrast, filmsetting has displaced hot-metal typesetting only to a very limited extent. In the present context, the importance of this distinction between technical and economic efficiency is important because whatever the labour-force consequences of a new process in a particular installation, its implications for the industry as a whole depend on the speed or otherwise with which it is taken up by firms. However dramatic or far-reaching the labour-force effects of a new innovation are expected to be, it is of little economic or industrial relations interest until firms actually begin to install it on a significant scale. What, therefore, are the factors stimulating or impeding the adoption of new techniques?

One of the most important is the trend and structure of a firm's output. We saw in Chapter 2 that the output of the printing industry grew at about 5 per cent a year between 1959 and 1967, and this has created the need for additional productive capacity. Firms have therefore installed new equipment, either of the same basic type as they were using previously, or which embodies some quite different technology. Even in the former case, the new equipment may still have a substantially greater capacity than earlier types, due to the incorporation of small improvements in successive models.

Although the growth in output has provided a favourable environment for the introduction of new equipment, we must still ask why this had led to the choice of one process rather than another. The principle motive for the introduction of both filmsetting and web-offset has been that the quality of the product is improved. This is particularly true of filmsetting, where the capital costs are high and labour savings are generally non-existent. But the quality of the type produced is much superior to that of hot-metal type in terms of clarity and definition, and for some types of work this outweighs the other disadvantages. Similarly, web-offset is a more expensive printing process than letterpress, but the product is more attractively printed and has a greater appeal to readers and advertisers.

The choice of other new techniques such as teletypesetting, powderless etching, sheet-fed rotary letterpress machines or the mechanization of the finishing process can be attributed to expectations of reduced labour costs sufficient to provide a satisfactory return on the investment. Particularly with powderless etching, a modest capital outlay (of about £1,200) enables equipment to be purchased which gives a very quick return if full use can be made of its potential benefits. Similarly, the introduction of TTS equipment roughly doubles a compositor's rate of output, though this saving is reduced by the employment of men to operate the tape-operated line-casting machines. We showed earlier that mechanized finishing equipment very substantially increases the productivity of the labour involved, allowing the numbers to be reduced or output to be increased. Finally, we have seen that sheet-fed rotary letterpress machines have a capital cost only a little greater than that of flat-bed machines, but that they reduce labour requirements per unit of output by two-thirds.

But while the value of these benefits will to some firms be sufficient to justify the capital expenditure, this will not necessarily be true of all firms in the industry. Other firms may decide that in their particular circumstances and production arrangements the investment would not be economic. One reason is that the structure of a firm's output may be such that it cannot get the full benefit of the technique, so that the capital cost is not worthwhile. A clear example of this is the use of TTS equipment by newspapers. This speeds up the setting of type, but the full benefit is only secured when large quantities of fairly straightforward setting are required. If a newspaper uses a number of different type-faces on a page, and frequently varies the column width, the benefits of TTS are lost. These editorial considerations will make the process much less suitable for popular newspapers than for those which produce substantial areas of standard type, such as provincial or 'quality' ones. Thus, at one plant which produces a morning and evening paper, TTS is used for the former but not for the latter, which is of a different editorial character. It is also technically possible to use TTS systems in a much more radical way in local newspapers. They receive much of their national and foreign material from news agencies and this could very easily be set centrally and linked by TTS systems to the composing rooms of provincial newspapers, with very substantial economies. But this is prevented by the present nature of the final product in two ways. First, a different importance and amount of space may be given to a news item by each editor. Secondly, the use of TTS in this way would require a standard column width in each paper and this degree of conformity does not exist.[1]

A similar problem has arisen in the case of mechanized publishing equipment where even if the equipment is installed by a firm, it may not be used for the whole of the relevant output. At one plant which was studied, mechanized equipment ties nearly all of the production, while at another company only about one-third of its output is handled in this way. The reason for the difference is that the equipment is most efficient when it is dealing with bundles of a uniform size. Non-standard bundles must be tied manually. For a large circulation paper this restriction presents few difficulties, as most orders from wholesalers are sufficiently large to be met by a number of standard bundles with perhaps one smaller one. These small packages have to be tied by hand, but they represent a negligible proportion of the total output. But at newspapers with a limited circulation a large proportion of the output will consist of bundles of widely varying sizes which cannot be handled by mechanized equipment at its present level of development. Therefore the effect of the equipment on the labour force is much more limited at the latter type of plant than at one with a larger and more standardized output.

The slow adoption of filmsetting can also be attributed to the view of many firms that the benefits are not sufficient in their plant to justify the considerable capital investment. Although it has often been introduced in association with a change to offset printing, mainly because of the higher quality obtained, many other firms in this situation have continued to set type in hot metal. One reason is the high capital cost. None of a firm's existing typesetting equipment can be adapted for filmsetting and must be discarded if the process is introduced. The processing units, and to a lesser extent the keyboards, are recent developments with an uncertain physical and economic life, and may be subject to rapid obsolescence. Although the expectation of continuing development has been the main reason for installing the equipment by firms establishing new plants, it has deterred others. In particular, a firm which already has hot-metal equipment may prefer to keep it for the time being in the hope that more efficient filmsetting equipment will be developed.

[1] It is understood, however, that at least one large chain of newspapers is gradually standardizing its column size and format so that it may eventually be able to make use of the system.

Furthermore, we noted earlier that there are no significant savings in manpower, with the result that the heavy capital costs are not offset by savings in operating costs. In part this is due to the cost of correction or alteration. These can be made very quickly and easily with hot-metal, but are much more time-consuming with film-set material. In newspapers this may cause an unacceptable delay in view of the tight production schedules to which many such firms operate. In other parts of the industry where time is less important it nonetheless increases cost, and one firm estimates that corrections to filmset work cost about five times as much as they do on work in hot metal. To overcome this another firm has instructed its key-board operators to work more slowly to reduce the cost of corrections. Considerations of this sort will naturally deter many firms from introducing filmsetting so that the rate at which it displaces hot-metal equipment is slow.

Finally, the use of web-offset in daily or evening newspapers has also been proceeding relatively slowly, though there are some notable and well publicized installations. One large evening news-paper which intended to replace its existing plant explored the alternatives in detail before deciding to invest in new letterpress machines, and its reasons are instructive. After examining the experience of a number of organizations in this country and in the USA, it calculated that both paper waste and ink costs would be roughly doubled, while the annual cost of plate production would rise from about £9,000 with current stereotypes to about £243,000 for offset plates. To this newspaper, however, the additional costs were less important than the loss of speed and flexibility. A paper of this type makes a number of changes in content and layout in successive editions, as well as adding stop-press news during editions. Even though offset plate-making has been simplified and the time taken reduced, it still lacked the speed and flexibility with which sets of stereotypes could be produced. The production pattern of a large newspaper is different from that of a smaller paper with less exacting schedules and this pattern depends on a degree of flexibility which web-offset has found difficult to provide.

The burden of the foregoing argument is then that although many new techniques have become available to the printing industry, it is important to recognize that this has not automatically ensured their wide adoption by the industry. In part this may be due to an inade-quate appreciation of their advantages, or a reluctance to adapt to new methods. But there are also some substantial economic reasons delaying the introduction of some types of new equipment, which are important to bear in mind when considering the labour-force effects of the new techniques.

III. EMPLOYMENT PROBLEMS

(i) *Labour Force Size and Occupational Structure*

There are several factors which complicate a simple comparison between the labour force required to operate a new and an old plant. One is that the rate of output on the new plant would almost certainly be greater than that which is being succeeded, and the labour-force requirements must obviously be standardized to take account of these differences. Secondly, the product may not be the same: one of the reasons for the introduction of the new process is to improve quality, and to standardize for this is practically impossible. More important, technological innovations are often introduced in a completely new establishment, and particularly if a new product is involved, there may be no displaced older plant with which to make a direct comparison. Finally, a very important complication, which we discuss in the next section, is that the bargaining over manning which is a feature of the industrial relations system of the industry may blur the exclusively technological cause of labour-force changes. Nonetheless, in the preceding chapters a fairly clear picture emerged of those new techniques which have substantial labour-force consequences and those which do not.

But in saying this it is important to recall that the extent to which a particular technique is able to save labour in any one plant may be greatly affected by the firm's output structure, so that the installation of the same new technique can have quite different labour-force effects in different firms. For example, several national newspapers have installed mechanized publishing equipment and, as shown above, the effects on employment have differed considerably, since the output of one firm enabled it to make full use of the equipment while the less standardized output of the other meant that it benefited rather less from the use of the equipment. Similarly, TTS can only be usefully applied when the work consists of fairly large quantities of material to be set in a standard format. If the newspaper, for editorial reasons, varies the column width or typeface frequently throughout the paper, TTS would save little or no labour

It is also interesting that the greatest effects on the industry's manpower have not necessarily come from the most sophisticated techniques. Neither filmsetting nor computer typesetting, both of which come from quite advanced scientific developments, have yet had very big consequences for the labour force as a whole, though they may affect certain specialist operations in a material way. Conversely, mechanized publishing or powderless etching which, although new techniques, are based on 'conventional science', have had a substantial effect on the labour force in firms where they have been introduced. We mentioned above that sophisticated pieces of equipment such as filmsetting units have a high capital cost but do not secure labour savings. Unless a firm is establishing a new plant or values the higher quality of production, and can persuade readers or advertisers to pay for it, there is little incentive to install filmsetting. In other words, although it is technically advanced, it is economic only in certain fairly restricted circumstances and similar considerations have affected the rate of introduction of computer typesetting.

The evidence from our cases can be discussed alongside the aggregate data on the printing industry labour force, and Table

TABLE 4.1

Numbers Employed and Percentage Change in Main Printing Occupations in 1964 and 1967

	Number Employed		
Occupation	1964	1967	*Percentage Change*
Machine Minders—Letterpress	22,980	22,680	−1·3
—Offset	4,780	5,200	+8·8
—Gravure	1,820	2,000	+9·9
Machine Assistants	17,810	17,780	−0·2
Compositors	48,910	46,800	−4·3
Process engraving workers	4,610	2,810	−39·1
Electrotypers and Stereotypers	4,640	3,830	−17·5
Skilled binding workers	25,240	29,060	+15·1
Semi-skilled binding workers	22,650	24,550	+8·4
Other manual workers	97,470	93,560	−4·1
All manual workers	250,910	248,270	−1·1

Source: Ministry of Labour *Gazette*, December 1964 and January 1968.

4.1 shows the numbers employed in the main occupations in 1964 and 1967.[1] Considering the occupations in the same order as we

[1] Other valuable series have been for many years collected by the British Federation of Master Printers and at least one of the trade unions, SLADE. However, it has been decided to use the official series as they have a wider coverage of firms than the BFMP. The latter series, and that of SLADE for the occupations in their membership, provide useful corroboration of the trends shown in the series presented.

did in the previous chapter, we can first see that there has been a slight decline in the number of letterpress machine minders, compared with a rise in the number of machine minders on offset and gravure work. This is broadly in line with what would be expected in the light of our earlier discussion. Although new printing equipment often has considerably greater capacity than older versions, this has generally been used to meet a growing product demand (or to reduce the time taken to print an edition in the newspaper industry) rather than to reduce the labour force. But most of this growth has occurred in the offset and gravure plants so that employment in letterpress is beginning to decline slightly.

In the preparatory trades we observed that the most serious labour-force problem had arisen from the introduction of powderless etching equipment into process engraving work. The effect of this on employment in this sector as a whole can be readily seen from Table 4.1. This is, however, almost certainly an overstatement of the decline, presumably due to a sampling problem.[1] However, examination of the alternative series prepared by SLADE confirms that there has been a distinct and real decline in the number employed over the period. In the composing operation a number of new techniques have been introduced with varying labour-force effects. The use of computers reduces the labour requirement of some complicated typesetting jobs by very large amounts, but if they are used only for justification work they save rather less than a quarter of the time taken for the work by a compositor. Furthermore, they have not been widely introduced. Filmsetting saves little labour as it is still necessary for the worker to operate a keyboard, and the effect in quantitative terms is not significant. But we saw that TTS equipment and even the more modern version of conventional hot-metal typesetting machinery raise labour productivity quite substantially, although again their introduction has been relatively gradual. Nonetheless, this does seem to be having an effect on the number of compositors employed which has declined slightly, even though the industry's output has risen.

Finally, we saw that substantial labour savings were achieved at some plants by the introduction of mechanical equipment at the finishing and packing stages. These operations are customarily labour intensive, and employ virtually all of the female manual labour in the industry. Although there has until recently been relatively little capital investment in these operations, much of the work appears to be

[1] The sample used covers only a small proportion of small firms, which are however, important in this part of the industry.

readily adaptable to the greater use of mechanical methods of moving materials, finishing, tying and packing. The previous chapter discussed several cases where the numbers required on particular jobs in both general printing and in national newspapers had been reduced, sometimes by as much as half. However, this is not reflected in the aggregate figures in Table 4.1 where both skilled and semi-skilled binding workers have increased in numbers. This trend is also seen in the BFMP series over a longer period; and it is presumably accounted for by the growing output of printed material to be handled. This, in fact, emerges in some of our cases where new plant was used to prevent a rise in the labour force rather than to reduce it.

(ii) *Changes in Skill Content*
We have so far been discussing changes in the numbers employed in particular occupations, but within some occupations technological change has removed all or part of the skill content of the job. For example, powderless etching equipment has removed much of the manual skill and judgement from etching work, as well as reducing the number of men required for a given volume of output. Similarly, the production of offset printing plates used to be a skilled job, but automatic equipment and pre-sensitized plates have reduced the operation to a routine level. Further, to the extent that a computer is used to assist typesetting, the skill and judgement which the compositor required to make correct decisions on justification and hyphenation are no longer necessary.

These operations have traditionally been performed by craft workers, and this position has not generally changed despite the reduction in skill requirements. Sometimes, as in process engraving or typesetting, the job titles and the union to which the operator belongs remain the same despite the changed skill requirements. It follows that the skill of those concerned is probably not being fully used. In other cases, of which offset plate-making is the best example, the 'de-skilled' work is done not by a skilled offset plate-maker (who would be a member of the ASLP) but by skilled men, particularly stereotypers who have been displaced from their traditional occupation. Further, at one firm the making up of type set in film is being done partly by traditional compositors but also by some displaced process engravers and stereotypers.

The effect of this is, of course, that a number of skilled men in the industry are not, in fact, engaged on skilled jobs. Although this increases a firm's costs by the amount by which the earnings of

93

skilled men exceeds that which would be paid to their less skilled replacements, the economic cost may not, in fact, be very great. We have seen that the number employed in each of the preparatory trades is declining and for process engravers in particular the prospects of alternative employment which makes use of their skills will often be remote. Consequently their displacement by, say, an unskilled man, would only have the effect of putting another man on to the labour market who, because his skill is now irrelevant, is of no greater value to the economy than the unskilled man. It may even be less if he is older and less mobile, because of his long-term attachment to the industry, than a less committed unskilled worker. An important proviso, however, is that recruitment and training to the declining trade is stopped or severely limited, otherwise considerable losses are incurred. In the case of the process engravers the numbers entering the trade have, in fact, been severely limited in the past few years as a result of trade union pressure, which suggests an incidental benefit of the apprenticeship ratios operated in the industry, in that they can be used to prevent uninformed entry to the trade.

(iii) *Adjustment Problems*
The problems associated with the large scale redundancy or re-deployment of labour have only rarely arisen in printing. There have, of course, been serious problems when newspapers have closed down, but these events are due to changes in the market conditions facing a particular newspaper and have not been the result of technological change within the industry. Part of the reason for there not having been severe quantitative adjustment problems is due to the nature of some new techniques themselves, in that they are not primarily designed as labour-saving equipment. For example, their main object may be to improve the quality of the product, and firms installing them do so for this reason, rather than to reduce their labour force. Furthermore, new techniques which *do* save labour have often been installed in conditions of rising demand. The effect in the firm concerned has thus been to prevent an increase in the labour force that would otherwise have taken place, rather than to reduce the level of employment. Thirdly, the true labour-force effects of new techniques have often been obliterated by agreements between firms and the trade unions, a problem to which we turn in the next section.

There have also been relatively few problems of qualitative adjustment, despite the marked changes in occupational structure discussed above. For example, it has been a comparatively simple process to retrain rotary letterpress minders to operate web-offset machines and, similarly, the firms which have installed sheet-fed rotary machines have quickly retrained flat-bed letterpress minders to their operation. The obvious explanation is that although to the outsider these machines appear complex and novel, to an intelligent machine minder with some years of experience in the industry there is little difficulty in moving between the machines, which in fact have a number of common features. Similarly, workers have moved fairly easily between hot-metal setting and filmsetting, where the only problem is the mastering of a new keyboard. Where the new keyboard is comparable to the old, as with the Monotype system, the adjustment has taken only a matter of days; but with other systems the adjustment has sometimes taken up to three months. Finally, we discussed above the de-skilling of several occupations; where, as in the case of offset plate-making, people previously engaged on other jobs have transferred to these duties, it follows that there have been no retraining problems worth mentioning.

(iv) *Effect on Wages*

The introduction of new techniques in printing has generally had the effect of increasing the wages of those employed on them. Partly this is due to the structure of wage rates in the industry, whereby these are higher on more complex machinery—for example, those employed on rotary letterpress machines are paid at a higher rate than those on flat-bed machines of comparable size. In addition, merit payments are widespread in the industry, and one would expect that a person with a mastery of a new skilled process as well as an old would be rewarded accordingly, in line with his enhanced value to other firms. For example, a firm which had installed filmsetting alongside its existing hot-metal equipment took the view that the men who had gone on to filmsetting were of greater value to the firm than those only skilled in hot-metal work, and paid a slightly higher merit payment. Similarly, a firm installing TTS equipment was paying, in 1967, approximately 30s a week more to the men on this equipment than to those on ordinary hot-metal machines.

In other cases, the effect of new techniques has been more complicated. One case, where the installation of new printing machines with a greater capacity threatened to reduce overtime earnings, was discussed in Chapter 3. The greater capacity enabled printing to start later in the evening, thus eliminating an overtime payment to

95

which the men had become accustomed. Recognizing this the firm proposed to make good the reduction by negotiating a reduction in the number of men required, and sharing the saving with the remaining employees. Eventually a settlement on these lines was achieved.

A similar negotiated reduction in staff and sharing of the savings has followed each introduction of mechanical publishing equipment in the national newspapers since 1963. The agreement with the trade union specifies that staffing on the new equipment shall be agreed, and then the savings which this represents over staff required for hand tying shall be divided equally between the firm and the remaining employees. In this agreement steps were also taken to even out fluctuations in earnings which varied with the size of the paper from week to week. A formula was agreed to base the current paging bonus on the paging of the previous year; this bonus would change each year in accordance with the number of pages produced, but would remain constant within that year.

Reflecting the pattern of wage settlement in the printing industry, which we discussed more fully in Chapter 2, most of the wage changes as a result of technological change have been determined at the individual plant rather than at national level. We noted one or two cases in the preceding paragraphs where national agreements lay down different rates for different types of machinery, so that if a firm introduces a new technique then at least part of any wage increase granted will be the result of the national agreement. But most of the wage effects of technological change in printing are the result of plant agreements, generally the result of negotiations between management and trade unions over the terms on which the new equipment would be introduced. In this industry manning of new equipment is generally a subject of negotiations rather than management prerogative, and obviously a firm will often have to make concessions on wages or other conditions if the trade union is to make concessions on manning which will allow the firm to secure the full benefits of the technological change.

In some cases the concession has taken the form of a straightforward wage increase, but more commonly it has been an agreed proportion of the savings from the new technique, as, for example, when mechanized publishing equipment was installed in some national newspapers. In other cases, however, the changes in remuneration have been more radical, leading, for example, to the provision of greater fringe benefits as part of the agreement which preceded the establishment of Southwark Offset in 1964, or to greater stability of earnings in the agreement made to cover the more efficient manning of mechanized publishing equipment. Both of these latter examples may be regarded as 'indirect' wage effects of technological change, in that while they followed the introduction of new equipment, they were not a *necessary* consequence. But instead of simply giving part of the savings in the form of increased weekly earnings, in both cases the opportunity was taken of the prevailing climate of change to introduce certain wider reforms of the payment structure.

IV. TRADE UNIONS AND TECHNOLOGICAL CHANGE

There is considerable published evidence relating to the inefficient use of the printing industry's labour force, and to the role of the trade unions in this respect. The 1962 Royal Commission on the Press took evidence from many witnesses in the newspaper and periodical industry, and also commissioned a short investigation from a firm of management consultants who concluded that about a third of the production staff employed by the national newspapers would not be required if a 'reasonable standard' of efficiency were applied.[1] The Commission concluded that 'the efficiency of the [national] newspaper industry falls far short of the ideal. Conditions in the provincial newspapers and the periodicals are better, but they too leave a good deal to be desired.'[2] More recently the Economist Intelligence Unit conducted a very detailed investigation of the national newspaper industry and concluded that in broad terms some £5 m annually could be saved by reducing the number of workers in the production departments to a more efficient level, without any change in methods or equipment. About three-quarters of these potential savings were found in the machine rooms, where it was considered that most of the companies examined could do without about half of their present labour force, yielding savings of some £3·5 m.[3] Meanwhile the National Board for Prices and Incomes took evidence from a number of companies in the general printing industry in connection with its investigation of a wage claim, and those interviewed estimated that an improvement in performance of between 10 per cent and 25 per cent could be achieved by a more efficient use of manpower. The Board consequently took the

---

[1] Royal Commission, op. cit., *Report*, Appendix XII.
[2] Ibid., para. 74.
[3] Economist Intelligence Unit, op. cit., Part I, p. 92.

view that there was ample scope in the industry for increasing earnings without increasing prices, by raising the efficiency of production.[1]

There can be little reasonable doubt, therefore, that trade union and management policies have led to an inefficient use of manpower in the printing industry. But in the present context we are concerned with a particular aspect of this problem, namely the effect which these policies have had on the introduction and effects of technological change, and in this respect the effects of trade union policies are much less clear cut than in regard to general manning levels, with which the reports we have cited were mainly concerned. For example, the Royal Commission began its discussion of production efficiency with an analysis of the industry's use of new plant and techniques, but discussed only two in any detail, TTS typesetting and mechanized publishing. And only in respect of the latter does it draw attention, correctly, to the delay in its introduction caused by trade union insistence on a 'no-redundancy' policy. It then went on to consider manning standards generally, and found these a much more serious matter, concluding that 'it is no doubt in manning standards that the greatest inefficiency occurs', taking the form of employing too many men, or paying them excessive 'extras', or employing them inefficiently on overtime.[2]

It has been necessary to emphasize this distinction between the effects of trade union policies on general manning standards of the sort mentioned in the previous paragraph, and their influence on the introduction and effects of technological change. We are primarily concerned with the latter problem, and in this respect we shall argue that although the trade unions have sometimes adopted policies towards new techniques which have reduced their efficiency, this has not always been the main cause of their slow rate of introduction. With some new techniques the effects of trade union policies on their efficiency of operation have been marginal, while in others the delay in adoption of the new technique can be attributed more to the general economics of the process itself in relation to existing processes, than to trade union policies.

There are three types of policy which a trade union wanting to protect its members' jobs can adopt towards technological change. It can delay an agreement for as long as possible so as to prevent the

1 National Board for Prices and Incomes, Report No. 2, *Wages, Costs and Prices in the Printing Industry* (Cmnd. 2750, HMSO: London, 1965), para. 49.
2 Royal Commission, op. cit., *Report*, para. 91.

equipment being introduced; it can negotiate manning standards which are in excess of some optimum or efficient level; and it can negotiate 'no-redundancy' agreements.

The clearest example of a long delay in reaching an agreement is that covering the introduction of mechanical publishing equipment. Mechanized publishing equipment appears to have been available (although in a less sophisticated form than current models) since 1956,[1] but agreement on its operation was not reached until 1960.[2] Under the terms of this agreement the firms guaranteed that no regular employee would be discharged due to mechanization. A formula was laid down which allowed staffings to be reduced as natural wastage of regular employees took effect. On the face of it this was a not unreasonable agreement, but two special factors make it appear rather less so. Natural wastage in the industry is very low; very few people actually leave the industry for other work, and as there is no fixed retirement age, many men continue to work until a very advanced age. Consequently, labour force adjustment would be a very slow process under this formula. But, in addition, certain categories of worker, closely but not directly involved in the packing operation, were excluded from the natural wastage calculations. If, for example, a man travelling on a van as a loader left for some reason, his place could not be taken by a man surplus to requirements in the packing section and the staffing there reduced accordingly—he had to be replaced by a new employee. Both these aspects of the 1960 Agreement made it very difficult for firms to secure the savings in labour costs which could be achieved by installing mechanized publishing equipment.

This position was remedied by a new agreement reached in 1963. This permitted staff to be cut to a more realistic level when mechanized equipment was introduced, provided the employment of existing regular employees, as distinct from casuals, was safeguarded. It also prepared the way for negotiations at plant level on decasualization of the publishing room labour force, by which firms would employ on a regular basis sufficient staff to handle an average-sized issue and this staff would also deal with loads above or below this level. Finally a formula was established by which the savings accruing either from mechanization or from decasualization would be shared between the firm and the remaining employees. This agreement was,

[1] Royal Commission, op. cit., *Documentary IV*, Evidence of the National Union of Printing, Bookbinding and Paper Workers, para. 14.
[2] Royal Commission, op. cit., *Oral I*, Evidence of the Newspaper Proprietors Association, particularly questions 1180–1183.

of course, between the trade union and the employers' association, and before any firm could introduce changes in a particular plant, it would have to negotiate with the union on the actual level of manning in that plant.

There are also a number of cases at particular plants where a union has made demands over staffing which a firm has been unwilling to meet, preferring not to use the equipment rather than accept what it regards as unnecessarily high manning. In other cases, of which the establishment of some new web-offset newspaper plants in 1967 by Thomson Newspapers was a well publicized example, a venture has been delayed or even threatened by a failure to agree on manning. But these isolated cases must be kept in perspective by setting them against the very large number of installations where an agreement has been reached quite quickly on the manning to be adopted, and it cannot be said that this particular union tactic has seriously hindered the introduction of technological change, with the exception of the delay of some four years before it was possible to install mechanized equipment in newspaper publishing rooms.

A second union policy is to negotiate agreements which lead to the employment of more labour, or labour of a more highly skilled or experienced type, than is necessary for efficient operation. Naturally there are conflicting views on the necessity or otherwise of a particular manning structure and all must be treated with care. Nonetheless, there are several cases where agreements have been reached on the employment of more men on a piece of new equipment than are required for its efficient operation. Two very clear examples occurred in relation to TTS equipment and powderless etching, where in both cases an agreement has been made specifying a particular level of manning which has later been replaced by one allowing a lower level. In the first case, agreement was made between the Scottish Daily Newspaper Society and the Scottish Typographical Association that each tape-operated casting machine should be supervised by one man. This contrasted to the comparable agreement in England which was that one man could operate two casting machines, and in 1968 a new agreement was signed bringing Scottish practice into line with that in England.

A similar train of events occurred when powderless etching was introduced and an agreement was reached to the effect that 'two men shall operate each machine',[1] but a later agreement removed this requirement and they can now be operated by 'any one etcher in

[1] Process Trade Agreement, January 14, 1960.

100

the course of his duties'.[1] But despite the union's efforts to mitigate the employment effects, powderless etching has entered the industry very rapidly. In some cases this is due to firms, particularly news-papers, wanting to take advantage of the greater speed of production which is possible with the equipment, rather than a reduction in labour costs being the principal motive. Other firms have put it in on the assumption that they will eventually be able to reduce their labour force, while others again are able to use the equipment to increase output with their existing labour force. In other words, the advantages of the machine are such that strong defensive action by the trade union has not succeeded in preventing its use by many firms in the industry.

The most serious problem which has arisen recently has concerned the manning of web-offset equipment. These presses are manned by a combination of craftsmen and assistants, and the question of the areas of work to be done by the two groups has caused difficulties. There were very few problems between the two craft unions, the NGA and the ASLP, prior to their amalgamation. A potential difficulty occurred if web-offset completely replaced letterpress plant, as has always been the case with its installation in existing newspapers. Here the existing labour force would belong to the NGA, while the new process had traditionally been handled by the ASLP. But there was little opposition by the latter union to letterpress minders taking over this work, though in some cases they have transferred their union membership from the NGA to the ASLP. For example at Southwark Offset the presses were operated by members of the NGA with only two ASLP men working in an advisory capacity as technicians. It seems that a shortage of litho minders (as a result of the control of entry to the trade) had enabled the ASLP to permit the letterpress minders to operate the process, especially in cases where to insist on their own members being employed would have led to the displacement of NGA men from their jobs.

A much more serious situation has arisen between the craft unions, particularly NGA, and the non-craft SOGAT. An illustration of the issues involved is provided by the dispute which occurred at South-wark Offset and which resulted in a Court of Inquiry being estab-lished by the Ministry of Labour.[2] In part, this dispute was over the level of staff to man the new equipment but the greater difficulty arose out of an attempt by SOGAT to extend its jurisdiction over work previously done by NGA members. The dispute over manning levels

[1] Process Trade Agreement, June 21, 1965.
[2] The Cameron Report, op. cit.

101

was essentially between the company and the NGA. The former had throughout the development emphasized its intention to break away from traditional practices in the area and obtain high levels of efficiency. In return for this the workers were offered substantial improvements in wages and other conditions. Agreement on the number of NGA men who would operate each size of press was reached with the NGA before the plant opened; for example, that a four-unit press would have two NGA members working on it. But as the plant came into operation, the NGA sought to re-open this question, claiming in particular that $3\frac{1}{2}$ men (later modified to 3) should be employed on each four-unit press. The company rejected this claim and eventually this dispute together with the further conflict with SOGAT came before the Court of Inquiry.[1]

A fundamental criticism made by the Court on this aspect of the dispute was of what it considered the inadequacy of the industry's arrangements for settling manning questions on new web-offset equipment. Because of the different characteristics of particular installations, it is impracticable to lay down national manning scales, and the report accepted that manning would normally have to be established at plant level, but it drew attention to what it considered the rule of thumb methods by which companies arrived at a proposed level of manning, which was, in turn, subject to pressures during bargaining with the trade unions. It argued that a more objective set of criteria for establishing manning in particular cases could be established, so that at least negotiations with the unions would have a more solid factual basis. The Court also had to deal with the problems arising from SOGAT's attempts to extend its jurisdiction into areas of work currently performed by the NGA, who had a monopoly of the senior jobs on the machines. SOGAT argued that no union had a prescriptive right to these new presses and that the matter was still to be resolved. It claimed that many of its members were well qualified for web-offset work, having had experience of it, in contrast to the NGA members who, in general, had none. Consequently, it saw no reason why they should not also be allowed retraining facilities to enable them to take over these presses.[2] Behind the SOGAT pressure for increased responsibility there may have been fear of declining job-opportunities. Web-offset presses in general printing reduce the labour force of minders and assistants, while greater mechanization of all forms of presses tends to reduce the requirements for assistants. In addition, there

[1] The Cameron Report, especially paras. 61–73.
[2] Ibid., para. 149.

is growing resentment at the lack of opportunities for their members for advancement to senior jobs.[1] One solution for the union would be if it were able to extend its jurisdiction to the more skilled minders' jobs.

The NGA, on the other hand, was obviously concerned to protect its members' jobs and was not prepared to give way. It argued that:

'the essential issue underlying the disputes which had arisen between itself and SOGAT at Manchester, Southwark and elsewhere was whether the jurisdiction which the ASLP and NGA had established for their members over the control of web-offset presses should be shared with SOGAT. In other words, the disputes had been caused by an overt act of aggression by SOGAT for which no justification had been offered. Neither ASLP nor the Association could afford to accept this claim which was being made by SOGAT, if for no other reason than that to do so would restrict the opportunities for employment available for their members while extending those available to members of that union.'[2]

The Court pointed out that the dispute fell within the scope of a rule of the Printing and Kindred Trades Federation which provides a procedure for settling such disputes, but that it was symptomatic of the attitudes of the unions that they were unwilling to submit their differences to this procedure. This reflects the strength of feeling over this matter and one cannot see a permanent solution to it while the two unions remain separate. Clearly SOGAT is likely to become more dissatisfied with the blocking of promotion prospects, while conversely the NGA is unlikely to be prepared to relax significantly their defensive role. In the immediate dispute the Court in effect recommended that the *status quo* should be maintained, and stated that it would deprecate any further attempt by SOGAT to extend its jurisdiction. The main reason for this seems to have been the hope that this would least endanger the possibility of a long-run solution being found by the merging of the two unions concerned. This does seem to be the only way out of the problem, given the present intransigence of the parties, but even this may not immediately eliminate conflict between sections within a merged union. These disputes naturally raise the question of whether they have materially delayed the introduction of web-offset into the industry. It should be emphasized, however, that many other web-offset plants have been installed without such difficulties,

1 Ibid., para. 150.
2 Ibid., para. 141.

albeit perhaps because managements have accepted inefficient working practices rather than risk the cost of disruption. But although these disputes have not helped, closer examination suggests that other reasons than union recalcitrance offer more substantial explanations of the apparent slowness with which the process has been adopted.

We have seen that the use of web-offset in newspapers does not reduce their cost of production and has generally only been installed in conditions of rising demand which have made existing printing equipment inadequate, or where a completely new paper is being established. Most provincial newspapers face little competition except at the fringes of their circulation area and in these circumstances, provided the existing letterpress equipment can satisfactorily produce the output, there is little incentive to start printing by the more expensive web-offset process. There may be an inducement in that advertising revenue will probably be increased if colour can be used, but this has to be offset against the additional capital costs which would be incurred. In addition, web-offset is at present ruled out for large newspapers because of the greater time and cost required for the production of duplicate plates compared with the production of stereotypes, and by the slower speed at which web-offset equipment prints compared with conventional rotary letterpress. These market and technical factors seem to be much more important in explaining the slow rate of adoption of web-offset printing by provincial newspapers than difficulties with the trade unions.

In general printing, the most important factor limiting the expansion of web-offset is probably the limited amount of work which it can produce economically. The high cost of preparation makes the process uneconomic below about 15,000 copies, while above 500,000 copies it is usually more expensive than gravure. It therefore depends on coloured work with an output between these two margins, and obviously there is only a limited, although growing, amount within this range. There is, in fact, some evidence of over-capacity in this part of the industry and much new web-offset equipment is said to be under-used, which in itself will discourage rapid adoption of the process. It was, however, notable that prior to the establishment of Southwark Offset there was no web-offset plant in Central London and this was generally attributed to the restrictive policies of the trade unions in the area. Nonetheless this had not prevented firms elsewhere in the country from establishing plants, and Southwark Offset itself was able to negotiate an agreement with the trade unions which promised to allow a satisfactory establishment in Central London.

Over-manning in a qualitative sense is also important. This occurs where, for example, skilled labour is employed on work from which much of the skill content has been removed. Again there are a number of instances of this occurring in printing. Probably the most indisputable is the normal practice to employ skilled compositors on making-up pages of filmset type. This is a job requiring considerable manual dexterity and seems well-suited to women. But firms have generally accepted the arguments and pressure from NGA that this is a job for the compositors, even though their previous experience has been in handling lumps of lead type which probably makes them less suitable for the job than women. That composing skills are not required is demonstrated by the experience at one plant which used some displaced process engravers for the work as well as compositors.

A similar case is the insistence of the NGA that only fully trained compositors should be allowed to operate filmsetting or type-setting keyboards. Although some filmsetting systems are similar to the earlier hot-metal casting machines, others are of a fundamentally different design. The keyboards of the latter are very similar to electric typewriters and employers have expressed the hope that they would be allowed either to use unskilled male labour or, preferably, female typists for this work: this, however, has been resisted by the NGA which has maintained the rights of its members to this work. The same applies to TTS tape keyboards, and even more strongly where a computer is to be used for justification work. In both, the traditional composing skills are largely removed and what remains would generally be within the capabilities of a competent typist.

We have seen that the introduction of filmsetting into the industry has been slow, and one would certainly expect that if firms were allowed to employ women on the work the use of the equipment would be increased. Because of their greater manual dexterity output would be high and therefore savings would be greater than if retrained compositors were used. The force of this argument would be increased if their wages were materially lower than those of men on the same work, though the importance of this consideration will probably diminish over the next few years. But one can still argue that the acceptance of women would not seriously affect the slow rate of introduction of filmsetting. As we discussed earlier, the fundamental reason for this is that the essentially manual nature of the operation remains, and in addition the cost of correcting filmset type is very much greater than that of type set in metal. This generally means that the savings in operating costs are not sufficient (or may not even exist) to make the substantial capital investment worth while, unless the firm and its customers value the higher quality of type set in this way, or are setting up a new offset printing plant anyway. While these disadvantages would be partly offset if firms were able to employ women at lower wages or higher levels of output they would only really influence those firms currently at the margin between filmsetting and hot-metal setting, and it seems unlikely that it would significantly increase the rate of adoption of filmsetting by the industry.

A third type of policy is that which secures a 'no-redundancy' agreement from firms introducing technological changes. These are also common in printing, being negotiated both nationally and on a plant basis. An example of the former was the mechanized publishing agreement of 1960 between the NPA and the London Central Branch of the NUPB & PW[1] which laid down that 'no regular man or regular casual shall be discharged due to mechanization'. The agreement contained a formula for calculating natural wastage; but this only applied to a list of specific occupations. If a man left an occupation not in the list, even though the job was of a similar character and was controlled by the same union, it was not possible to transfer a man from the section with a surplus to the section with a shortage: a new employee had to be taken on and this obviously reduced the rate at which the surplus of redundant men was reduced. Given, therefore, that savings would accrue very slowly, most firms whose output structure was suitable were nonetheless unwilling to invest in the equipment. This position was changed by the 1963 agreement which did permit firms to adjust their labour force, enabling benefits to be secured from the investment and most national newspapers have now installed or are installing the equipment. But it is also worth recalling the point we made earlier, that the technique is only economic for large circulation newspapers, most of whose output is distributed in standard-size bundles: papers with a smaller circulation and fewer standard bundles can use it for only a proportion of their output and it is often not worth installing it on that basis.

A curious anomaly arose at one national newspaper which was re-negotiating manning levels having installed new printing machines. It made an agreement to the effect that although the number of

1 Memorandum of Agreement between The Newspaper Proprietors Association and The National Union of Printing, Bookbinding and Paper Workers (London Central Branch), March 8, 1960.

casual workers would be reduced, no regular employees should be made redundant. As proportionally more regular employees were employed on the Sunday paper than on the daily, the effect of having to provide jobs for them all meant that when the machines were printing the Sunday paper they were manned by $18\frac{1}{2}$ men, while for the rest of the week there were only 15 men to each machine.

SLADE has negotiated a number of no-redundancy agreements at plant level. As a result, for example, one firm installed powderless etching equipment and estimated that it could readily reduce the number of process engravers from ten to seven, but was prevented by the agreement from doing so. It also found that the union then attempted to extend the interpretation from protection of existing holders of a job to protection of the job itself. One of the ten men eventually left, and instead of allowing the job to lapse, the firm came under severe pressure from the trade union to take on a replacement. This was resisted for several months, but eventually the firm gave way and employed another man.

The basis of union pressures for no-redundancy agreements which seem to have been very widespread in printing is, of course, that the workers have secured a form of property right in their jobs. In some cases these 'rights' can be very valuable, if it is considered that a man's earnings in some alternative job are much less than his present one. This view was clearly expressed by the NUPB & PW in their evidence to the Royal Commission on the Press when discussing the delay in reaching agreement on mechanized publishing equipment. The work in the publishing rooms is at most semi-skilled and is also specific to this industry, so that the earnings of the men were undoubtedly very much higher than they would have received in alternative occupations or industries. Consequently the men and the union were prepared to use their strength in negotiation to protect these jobs, and this they successfully did for several years, undoubtedly discouraging firms from spending the considerable sums in return for very little benefit until the labour force eventually fell through natural wastage.

The cost of policies of this sort largely depends on how much surplus labour there is, and how quickly it leaves the firm. Turnover in printing is in any case lower than the average for all industries,[1] but in addition if restrictive conditions of the sort mentioned from

---

[1] For example, in one recent 4-week period the number of discharges and other losses in the newspaper and periodicals industry was 1·7 per 100 employees, and was 2·1 in general printing. The comparable figure for all manufacturing industry was 2·6. Ministry of Labour *Gazette*, July 1968, p. 556.

the mechanized publishing agreement are applied, the rate of natural wastage may be very low indeed. Furthermore, the craft-based structure of a firm's labour force will severely limit the amount of internal transfer which can be made to absorb the surplus labour. Even this can be mitigated by careful forward planning, as at one firm which was installing new binding equipment which severely reduced its labour requirements in a particular section. Because of the difficulties of internal transfer it used the comparatively long period between the decision to order the equipment and its eventual installation to run down the labour force on the section so that by the time the new equipment arrived the labour force approximated to the required level.

The question then arises of how serious a deterrent to new investment no-redundancy policies have been in printing. Clearly it delayed the introduction of mechanized publishing equipment into the national newspapers, but other new techniques appear to have been much less affected. The strongest reason for this has been that many new techniques have been installed in response to a rising demand for the product which in general terms it has then been possible to meet with the existing labour force: the point has been to prevent an increase in the labour force rather than to reduce it. In other cases, even though a firm has had to carry surplus labour, this disadvantage has been outweighed by the other advantages of the new technique, such as the improved quality of the product, or greater speed of production, and little evidence emerged, apart from that of mechanized publishing, where the introduction of new equipment had been seriously hampered by no-redundancy policies. The main danger of no-redundancy policies is that once a firm has accepted it, and established a manning level on new plant which contains an element of surplus labour, it may find it difficult to remove this later on. There is evidence both in this industry and in others that once a particular manning level and a set of informal output norms have been established it can be very difficult to revise them at a later date, with the result that some permanent over-manning can be built into the operation, a problem to which we return in Chapter 11.

Finally, there is the problem of whether inter-union demarcation disputes have seriously affected the introduction of new techniques. It has been commonplace for many years to draw attention to the complex structure of trade union organization in British industry generally, and the printing industry has often been cited as a particularly bad example. It has been argued that the rigid inter-craft demarcations have created serious inefficiencies and are at the root of much of the industry's wasteful use of labour. This is not the place to discuss inter-union demarcation issues generally, but we can now say something about it in relation to technological change. From the earlier discussion it emerges that in only two areas have serious difficulties arisen between two trade unions in recent years. These were the dispute between SOGAT and the NGA over their respective spheres of influence, and a fairly minor one between NGA and SLADE.[1] In contrast, the ASLP has not prevented NGA members from manning offset presses, or stereotypers (belonging to the Amalgamated Society of Electrotypers and Stereotypers) from making offset printing plates where they would otherwise have been made redundant. It is a matter of judgement whether the benefits of these arrangements outweigh the losses from the NGA/SOGAT disputes, but in general inter-union difficulties have not been a major problem of technological change in the industry.

In conclusion, it appears that although the trade unions have frequently secured agreements with the employers which have the effect of reducing the rate of return on the investment in new techniques, the importance of this factor in explaining the rate of introduction of new equipment should not be overstated. Obviously there will always be firms where the decision to invest or not in a particular new machine is finely balanced, and where union policies result in tipping the balance against the investment. But only the installation of mechanized publishing in national newspapers appears to have been generally delayed by union policies. Elsewhere other factors have been more important than union policies in delaying the adoption of new equipment or else the equipment has been installed in spite of these policies. On the other hand, this does not alter the earlier conclusion that agreements have been made which reduce the efficiency of new investment. While this may often have only deterred marginal installation it does mean that resources are being wasted in the firms who have followed the terms of agreements of this sort, and that profits are lower or prices higher than they would otherwise have been.

[1] A problem arose many years ago between the National Union of Press Telegraphists and the predecessors of the National Graphical Association, but this was resolved satisfactorily.

PART THREE

THE STEEL INDUSTRY

CHAPTER 5

THE INDUSTRIAL BACKGROUND

I. INTRODUCTION

Since the 1930s the British steel industry has undergone considerable changes of fortune. During the immediate pre-war years the industry was still suffering from depression and was in need of considerable modernization and rationalization. Between 1939 and 1945 longer-term plans were largely shelved, the requirements of the war economy being paramount. After 1945 the industry embarked upon a programme of modernization and expansion of capacity, but now found itself hampered by controls and shortages of raw materials. Only in the 1950s did the industry begin to make real progress towards rationalization and planned development, although some controls (for example on exports) and shortages persisted. Meanwhile, home demand for steel was continuing to rise and there was an urgent need to expand capacity.

It is probably true to say that capacity caught up with demand only around 1960. But by 1964 a serious surplus capacity problem was developing in world steel production, though this did not apply equally to all products. The international market became increasingly competitive, with many countries exporting at prices below the level of their domestic prices. Exports from Britain were running at 15 to 20 per cent of tonnage output, but the problem of maintaining this level at profitable prices was growing increasingly severe. The appearance of surplus capacity also had implications for imports, although the full effects of this were moderated for a time by the import surcharge imposed in October 1964. The market for steel is thus no longer a seller's market as it was between 1945 and 1963.[1]

The implications of this reversal in market circumstances for

[1] For a comprehensive treatment of the historical development of the industry, see D. L. Burn, *The Economic History of Steelmaking, 1867–1939*, Cambridge University Press, Cambridge, 1940; and D. L. Burn, *The Steel Industry 1939–1959*, Cambridge University Press, Cambridge, 1961.

costs, investment and technological change are obvious. The British industry needs to sustain a high level of exports to keep its existing plant working at levels of capacity utilization high enough to yield competitive unit costs, but in order to be able to compete in the world market it must continually seek to reduce unit costs still further by investing in new, improved methods of production. Since advances in steel technology have been rapid and tend to be quickly disseminated among steel-producing nations—especially those engaging in world trade—the need for the British industry to keep abreast with, if not ahead of, its competitors in applying technological change is reinforced.

Until 1967, the industry faced up to these problems as a group of private companies, though subject to some degree of central regulation. After a very brief period of nationalization, the companies were mainly returned to private ownership in 1953, and the Iron and Steel Board was then established, with powers to control major investment decisions and pricing. In 1967 the major steel producing companies, accounting for about 90 per cent of crude steel output, were renationalized, and considerable organizational changes have been occurring and seem likely to continue. This does not alter the basic problems of the industry though it may affect the way in which it sets about tackling them, which raises a question that must be considered at the outset: how far does the change in ownership affect the conclusions drawn in the subsequent chapters, based as they are on trends in technological change and experience of labour force adjustment during the pre-nationalization era, mainly between 1961 and 1966?

There are several reasons why the present discussion is likely to have a good deal of bearing on the future activities of the industry.

(i) The course of the industry in the early phase of nationalization is likely to be heavily influenced by the investment programme of the last few years under private ownership. Only in the second half of the 1970s is it likely that major changes, such as new greenfield-site steelworks, will come into effect.

(ii) Secondly, the nature of technological developments over the next few years is not likely to differ greatly from that of the changes discussed in Chapter 6. Rather than rapid development and application of totally new technological concepts, it is more probable that there will be an extension in the use of technology that has been recently developed and still only partially applied. In so far as this is the case, the experience of implementing change in the past will be a useful guide to the problems of the immediate future.

(iii) There has inevitably been a diversity of approach to the problems of adjusting to technological change in the past, reflecting the multiplicity of companies each with its own special problems and its own policies on adjustment. The British Steel Corporation has explicitly recognized the need to weld together the previously separate organizations and 'to assimilate the wide variety of the former companies' practices to the level of the best'.[1] Providing that the forecast of (ii) above is reasonably accurate, the examination of difficulties experienced by companies in the past and their methods of approach to the implementation of change may well have a direct relevance to the formulation of an improved set of policies on a more uniform basis.

Overall, then, it is arguable that although the organization and structure of the industry have changed, the essential problems of technological advance are likely to remain much the same for some time to come, and a knowledge of the past may be useful in establishing policies for the future. On that assumption, the background set out in this chapter deals with the industry as it was organized prior to renationalization. Section II defines the industry as it is treated in this book and identifies the main product groups. Section III discusses the structure of the industry and recent trends in output and investment. In Section IV the organization and importance of research and development are summarized, while Sections V and VI provide some background on labour force trends and industrial relations respectively. The chapter ends with a brief summary of the main points.

II. THE INDUSTRY AND ITS PRODUCTS

In the Standard Industrial Classification of 1958, the iron and steel industry comprises two minimum list headings: 311 (Iron and Steel, General) and 312 (Steel Tubes). Iron-casting and non-ferrous metal productions are excluded. However, the Department of Employment and Productivity figures for employment under these two headings are based on the major activity of the establishment making the return and in some establishments there is also considerable employment in engineering activities, which is included in the iron and steel industry. A more precise approach is that which

[1] British Steel Corporation, *Annual Report and Accounts*, HMSO: London, 1969, p. 27.

was adopted by the Iron and Steel Board and the British Iron and Steel Federation (BISF) who jointly collected and published one of the most comprehensive sets of industrial statistics available in Britain. Under this definition[1], those parts of establishments not concerned with the manufacture of iron for steel-making, and making of steel and its working and treatment, were excluded. Since for the period of our study these were the official industry figures we will confine attention to them wherever possible.

The industry comprises three main production operations: the making of iron, the production of crude steel from iron and scrap and the shaping of crude steel into a wide range of semi-finished and finished products.[2] The industry's output can be divided into

[1] The steel industry, as covered by the iron and steel statistics published by the Board and the Federation relating to production and deliveries of iron and steel products, was defined in the Third Schedule to the 1953 Iron and Steel Act as follows:

'1. The quarrying or mining of iron-ore or the treatment of preparation of iron-ore smelting.

2. The smelting of iron-ore in a blast furnace with or without other metalliferrous meterials, or the production of iron by any other process.

3. The production of steel by any process.

4. The casting of iron or steel by any process.

5. The rolling, with or without heat, of any iron and steel products for the purpose of reducing the cross-sectional area thereof.

6. The production, with or without heat, of iron or steel forgings, but not including:

  (a) smiths' hand forging;

  (b) the production of bolts, nuts, screws, rivets or springs;

  (c) drop forging or any other stamping or pressing involving the use either of a die conforming to the shape of the final product of the stamping or pressing, or of a series of dies, one of which conforms;

  (d) the hammering or pressing of any part or component of plant or machinery carried out incidentally to, and by the persons engaged in, the manufacture or repair of the plant or machinery in which the part or component is to be incorporated.

7. The production from iron or steel of bright bars or of hot-finished tubes or of hot-finished pipes.

8. The production of tinplate or terneplate.

The production of pig iron and the production of steel in the form of ingots, slabs, blooms or billets shall be deemed not to fall within paragraph 4 of this Schedule but to fall within paragraph 2 or paragraph 3 thereof, as the case may be.'

[2] 'Finished steel products' has a special meaning in this context. These products are generally sold to other industries for further manufacture and treatment and are in that sense intermediate products. But they are finished products so far as the steel-producing firms are concerned. We should also note here that the

116

'common steel', and 'alloy and special steels', the latter amounting only to about 7 per cent of the industry's output by volume in 1965. Among common steels, which are our main concern here, normal practice is to distinguish three main product groups:

(a) *heavy steel products*, including heavy plates, heavy sections and bars, joists and heavy rails;
(b) *thin flat products*, including sheet, tin-plate and light plates;
(c) *light rolled products*, including light sections and bars, light rails, narrow strip and wire rods.

These three groups of products accounted for over 87 per cent of gross deliveries of finished steel products from UK production in 1965. The remaining group of common steel products comprises mainly tubes, pipes and fittings, and some types of forgings, railway tyres, wheels and axles.

III. STRUCTURE, OUTPUT AND INVESTMENT

The number of companies engaging in steel-making is not large in simple numerical terms. According to a BISF report there were 20 such companies in 1965, 8 of which contributed more than 70 per cent of total national capacity.[1] But the technical economies which are so important are not generally to be found at company level but rather in the individual works. It was common for a single company to have several works widely distributed over the country so that there was no possibility of operating them in a completely integrated fashion. As Table 5.1 shows, in 1966 there were 285 works whose principal activity was the manufacture of *common steel* products, but by no means all of those were engaged in steel-making. Works engaging in each of the three main production stages (iron-making, steel-making and finishing) are termed *integrated* works. *Non-integrated* works have no iron-making facilities while *re-rolling* works make neither iron nor steel, but engage solely in rolling steel independently or in association with a steel-producing

---

heterogeneity of steel products leads to some complications in measurement. On average, every 100 tons of finished steel requires 137 tons of crude steel, so that crude steel production exceeds the output of finished steel products by over one-third. For this reason, a standard measure of steel output is commonly used, the *ingot ton*, which is the unit of crude steel weight equivalent to a given quantity of semi-finished and finished steel: i.e. the equivalent of 100 tons of finished steel would be 137 ingot tons, on average.
[1] *Steel Review*, 1967.

TABLE 5.1

Analysis of Works Producing Common Steel Products, 1966

Type of works	Number
1. Integrated	21
2. Non-integrated	13
3. Iron-making only	7
4. Re-rolling works allied to steel-making works	28
5. Independent re-rolling works	70
6. Other specialized works	146
Total	285

Source: *The Steel Industry* (Stage 1 Report), BISF, 1966

works. There are in addition a large number of highly specialized works. Table 5.1 shows 34 works in the integrated and non-integrated categories in 1966—and it is these that are our main concern in this study, since they produce over 90 per cent of crude steel and provide the bulk of employment, as well as being the primary vehicles of major technological change.[1]

Table 5.2 indicates the size distribution of these 34 works, showing

TABLE 5.2

Analysis of Steel-making Works by Size, 1966

Annual capacity	Number
Under ½ million tons	15
½ to 1 million tons	7
1 to 2 million tons	9
2 to 3 million tons	2
3 to 4 million tons	1
Total	34

Source: *The Steel Industry* (Stage 1 Report), BISF, 1966.

that only three works had a capacity of over two million tons: they accounted for about 23 per cent of total national capacity, while the 22 works with less than one million tons capacity accounted for about 40 per cent of total capacity.

While the British industry did not compare altogether unfavourably with other major steel-producing countries in the size of the 'tail' of small works, it lacked really large scale plants such as existed in the USA, Japan, the USSR and to a lesser extent the ECSC

[1] In addition, there were 63 works engaged mainly in the production of stainless and alloy steels, only one of which was integrated: *The Steel Industry*, Stage 1 Report of the Development Coordinating Committee of the BISF, July 1966, p. 92.

118

countries. And this, as we shall see, was the situation at the end of a period of heavy capacity investment in the British industry.

Throughout the post-war period, the trend of crude steel output has been steadily upward, rising from just under 12 million tons in 1945 to 27 million tons in 1965. Between 1957 and 1965, which can be regarded as broadly comparable years in terms of capacity utilization and general market conditions, output rose by 24 per cent, an average of 3 per cent per year. Thus steel, though not one of the economy's fastest growing industries, has maintained a fair growth record. The trend was, however, interrupted by recessions in 1952, 1958, 1962–3 and 1966–7. Because steel is a basic industry, dependent on demand conditions in a wide variety of manufacturing and other industries, it is very sensitive to changes in the level of business activity. Yet it also is an industry of major national importance, with some obligation to provide adequate capacity to meet peak demands. Until about 1960, as mentioned above, capacity was still trying to catch up with demand, but once that objective was achieved the need for capacity to expand ahead of demand became important, and much of the investment of the early 1960s was aimed at providing adequate capacity for 1970. In these circumstances, however, there is always a risk—which became reality in 1962 and 1966—that the level of capacity utilization will decline sharply at times.

TABLE 5.3

Capital Expenditure, 1957–1966

Year	Capital expenditure (£m)	Capital expenditure index (1960 = 100)
1957	95	65
1958	105	72
1959	99	68
1960	146	100
1961	199	136
1962	170	117
1963	77	53
1964	55	38
1965	50	34
1966	42	29

Source: Derived from Iron and Steel Board Annual *Reports*, 1962 and 1966.

Table 5.3 shows the levels of capital expenditure in the period 1957–66.[1] Investment, though obviously erratic, was not unstable in

[1] This refers only to expenditure on capital developments costing over £100,000, and thus includes all major investments but is not quite comprehensive.

119

the same sense as this might be used to describe capital formation in the chemical industry. The period 1957–62 was one of major development in the industry, for the reasons already mentioned. Once the longer-term need for additional capacity had been met, investment declined sharply and it is likely that until the early 1970s there will be no return to the investment levels achieved in 1960–2, mainly because present capacity is now more than adequate for projected demand until 1975[1], leaving as the principal interim objective the removal of high cost, obsolescent plant and (following renationalization) a regrouping of works.

In Table 5.4 the allocation of investment among production

TABLE 5.4

Capital Expenditure by Main Process Stages, 1960–6:
Actual Expenditure (£m) and Percentage Distribution

Process	Coke Ovens	Raw Materials Handling and Blast Furnaces	Melting Shops	Mills and Finishing	Unclassified
Year					
1960	£8·5m	£30·1m	£18·6m	£83·6m	£5·3m
	(5·8%)	(20·6%)	(12·7%)	(57·3%)	(3·6%)
1961	£9·9m	£39·9m	£31·6m	£112·6m	£4·8m
	(5·0%)	(20·1%)	(15·9%)	(56·5%)	(2·4%)
1962	£9·6m	£30·8m	£32·4m	£92·7m	£4·7m
	(5·6%)	(18·1%)	(19·0%)	(54·5%)	(2·8%)
1963	£1·3m	£7·8m	£21·9m	£41·6m	£4·2m
	(1·7%)	(10·1%)	(28·5%)	(54·2%)	(5·5%)
1964	£0·8m	£6·5m	£15·3m	£28·9m	£3·4m
	(1·4%)	(11·9%)	(27·9%)	(52·6%)	(6·2%)
1965	£0·6m	£7·7m	£8·5m	£28·9m	£3·9m
	(1·2%)	(15·5%)	(17·1%)	(58·3%)	(7·9%)
1966	£0·4m	£3·9m	£6·5m	£26·9m	£4·0m
	(1.0%)	(9·3%)	(15·6%)	(64·5%)	(9·6%)

Source: Derived from Iron and Steel Board Annual *Reports*, 1965 and 1966.

stages and processes is shown for the years 1960 to 1966. Over one-half of total investment has consistently gone to the rolling mill and finishing stages of production, and the variations in the proportion of the total in any year going to these stages have been fairly small, though of course actual expenditure varied enormously. But in 1960–2 there was a relative expansion of investment in the

[1] This follows from the conclusions of the *Report* (Stage 1) of the Benson Committee in 1966, which by basing its demand on the 1965 National Plan assumptions, certainly overestimated capacity needs. Cf. *The Steel Industry* (op. cit.).

primary stages (coke-ovens, raw materials handling and blast furnaces and ancillary equipment); while the investment in steel-making (melting shops) increased relative to the total in 1962–4. The marked peak of expenditure as a whole in 1960–2 presents an interesting example of the bunching of investment projects which is characteristic of the industry. This was not simply due to the cluster-ing of capital projects *within a plant*, such as arises out of the need to undertake expansion at various production stages in an integrated fashion. A major explanatory factor was the industry's belief that it needed to effect a major expansion in capacity, resulting in a programme of development including the construction of new greenfield-site steelworks which inevitably boosted investment to a new high level.

This view was expressed through the Iron and Steel Board, the main organ of central regulation for the industry from 1953 to 1967. One of the main duties of the Board was to undertake a regular review of capacity, and by consultation with individual firms and trade associations (in practice, mainly the British Iron and Steel Federation) to ensure that capacity and the supply of raw materials were available for 'efficient, economic and adequate supply' of steel. Gaps in development could be reported to the appropriate Minister, who could take positive steps to right the imbalance. Any major development plan (effectively those costing over £100,000) could be vetoed by the Board if it did not meet the above criteria. The Board was also to be advised of closures of existing capacity. In practice the bulk of capital expenditure was concerned with the replacement of plant at the end of its physical life, and with the expansion of capacity. Few cases, if any, involved the replacement of major plant items purely because of obsolescence.[1]

Pricing policy was also influenced by the Board, and this too had a bearing on investment. The Board exercised its discretionary power to fix maximum prices over a wide range of products. Being statutorily required to ensure that prices were consistent with the development of an efficient industry, the Board undertook periodic reviews based on investigations of cost changes 'at modern plant operating under optimum conditions'.[2] In view of this, and the fact that the maximum prices set by the Board were also the effective

[1] However, the 1966 Annual *Report* of the Iron and Steel Board noted that there had been, in 1966, 'a more marked tendency on the part of steel-makers to discard even relatively modern open-hearth or other steel-making capacity and to replace it by LD vessels' (loc. cit., p. 54).
[2] Annual *Report*, Iron and Steel Board, 1965, p. 3.

121

prices until 1965,[1] the Board's power over prices should have meant reasonable profits, adequate for the financing of new investment and the incorporation of technological progress. The Board's additional power of veto on investment planning, and its ability to consult steel producers and users, meant that there was scope for eliminating wasteful duplication of investment and inadequate or technologically backward development schemes. From this standpoint, the climate for technological change in the industry was reasonably favourable, but the Board occasionally complained that its power was too negative and that its statutory functions were a reflection of the scarcity conditions prevalent when the Board was established, rather than of the world surplus situation of the mid-1960s. For example, in 1965 the Board expressed the view that its powers had

'sufficed to bring about the necessary expansion of capacity and kept prices at very reasonable levels, but they had not proved to be as effective as was hoped in stimulating efficiency nor can they be used to force the industry to adapt itself to present-day requirements.'[2]

One implication of this reference to 'adaptation to present-day requirements' is almost certainly the need for the industry to take better advantage of the very considerable technical economies of scale which exist in the industry. The statistics on the industry's structure presented at the beginning of this section are a clear indication of the inability of the Board to secure an organization capable of achieving these economies, however favourable its performance may have been in other respects.

IV. RESEARCH AND DEVELOPMENT

The British steel industry's expenditure on R and D is not necessarily an accurate measure of its technological performance, for in this industry many innovations are international, through licensing arrangements, and some of the most significant changes discussed below originated abroad. Nevertheless it is important to know what resources the industry has allocated to R and D, and how it has used them. Prior to renationalization the bulk of the expenditure was undertaken by the individual companies. This was supplemented

[1] Due to a number of trading agreements eliminating price competition: these ended in 1965 following the adverse judgement on the Heavy Steel Agreement in the Restrictive Practices Court.
[2] Annual *Report* (op. cit.), p. 4.

by the work of a central research organization, the British Iron and Steel Association (BISRA), which still exists. At the time BISRA was financed mainly by the companies through the British Iron and Steel Federation but also received a government grant (about £250,000 in 1966).

Table 5.5 shows the broad allocation of expenditure between BISRA and the Companies' own research organisations. BISRA

TABLE 5.5

Research and Development Expenditure in the Iron and Steel Industry, 1954–5 and 1962 to 1965 (£m)

	1954–5	1962	1963	1964	1965
Member Companies	3·0	6·31	8·02	10·71	10·06
BISRA	0·5	1·06	1·22	1·24	1·46
Total	3·5	7·37	9·24	11·95	11·52
Total at 1965 prices	4·9	8·2	10·0	12·5	11·52

Source: 'The Steel Industry' (op. cit.), Appendix 4.

accounted for about 12 per cent of total expenditure in the early 1960s. The Table also shows that in the period 1955–65 expenditure was increasing, and that in real terms expenditure had more than doubled.

Information on how the iron and steel industry compares with other industries is not completely satisfactory, but Board of Trade estimates suggest that in 1964–5 the industry accounted for about 2·4 per cent of all R and D expenditure in manufacturing industry. In this respect it is greatly inferior to chemicals but its share of the total was probably increasing in the early 1960s, since a Board of Trade estimate for 1961–2 gave iron and steel only about 1·5 per cent of the manufacturing industry total.[1] This relative rise may in part be explicable in terms of the view expressed by the Iron and Steel Board that 1 per cent of turnover was a reasonable approximate measure of the total research expenditure at which the industry should aim.[2] Estimates by the Board showed that in the 1950s the industry's research expenditure was less than 0·5 per cent of gross output, though by 1962–3 this had risen to 0·86 and in 1965–6 to 0·91 per cent.[3] Thus the target set in 1963 had not been attained by 1966. Turnover, of course, may vary slightly from year to year, depending on the demand conditions of the industry, and in fact

[1] The Board of Trade estimates are derived from sample inquiries.
[2] *Research in the Iron and Steel Industry, Special Report*, 1963.
[3] Iron and Steel Board, Annual *Report*, 1966, p. 51.

output in 1966 was down on previous years, so that the real situation was rather worse in relation to the Board's target than the figure of 0·91 per cent would suggest.

The division of the industry's expenditure on R and D into basic and applied research and development is difficult to make, but the Board of Trade source already quoted provides estimates of the distribution for the companies. About 5 per cent of current expenditure went to basic research in 1964–5, 40 per cent to applied research, and the remainder to development. Thus the emphasis at company level was very definitely on applied research and development. (No comparable figures are available for the work done by BISRA.) However, although roughly the same proportion of total R and D current expenditure by the steel companies went to applied research and to development as in manufacturing industry as a whole, the balance was rather different: for whereas 40 per cent of this went to applied research in steel, the all-manufacturing average was only about 20 per cent.

The actual areas of research and development are wide-ranging, covering product development, mechanical and plant engineering and the development of new methods of iron- and steel-making. Individual companies took an interest in all these areas, though at least in the larger companies the longer-term work tended to be undertaken largely by the company's own central research laboratories while the day-to-day problems of meeting the specific needs of particular customer requirements were met in works' laboratories. The role of BISRA was two-fold: firstly, to undertake research of common interest to member companies of the Federation, and secondly, to investigate certain problems referred to them by companies on a contractual basis.

The industry, therefore, is one which has a well-established interest and organizational framework for research and development; and while the intensity of investment in R and D does not compare with the chemical industry, steel does figure among the leading industries in this respect. From this have come a number of important technological developments, although the major technological changes in steel-making in the last two decades have come from abroad.

V. LABOUR FORCE TRENDS

The iron and steel labour force is conventionally classified into three broad occupational categories. First, there are the *process workers,* directly engaged in iron and steel production, treatment and finishing. Many of these are highly skilled, although no formal craft apprenticeship system exists. These skilled workers normally have most seniority in the plant, and below them are progressively less skilled workers with shorter service until we come to the labourers with least seniority of all. Secondly, there is the rather heterogeneous group of *general and maintenance workers,* including skilled craftsmen (e.g. bricklayers, electricians, etc.) as well as general labourers and all workers in the Service Departments (e.g. transport, canteens, etc.). Thirdly, the *administrative, technical and clerical* group includes management, as well as a range of other white-collar workers including scientists, professional engineers, technicians in laboratories and the clerical grades.

Table 5.6 shows the changing size and importance of these three groups for selected years between 1957 and 1966. However, because

TABLE 5.6

Changes in the Composition of Numbers at Work in the Iron and Steel Industry (excluding iron-foundries): selected years 1957–66

	(1)	(2)		(3)		(4)	
		Process Workers		General and Maintenance		Admin. Tech. & Clerical	
	Total Nos.						
	at Work	Nos.	As %	Nos.	As %	Nos.	As %
Year	(000s)	(000s)	of (1)	(000s)	of (1)	(000s)	of (1)
1957	311·0	166·7	53·6	98·0	31·5	46·2	14·9
1959	298·6	153·0	51·2	97·7	32·7	47·9	16·1
1961	306·9	153·9	50·1	100·2	32·7	52·7	17·2
1963	294·9	142·9	48·5	97·0	32·9	55·0	18·6
1964	304·9	147·8	48·5	99·2	32·5	57·9	19·0
1965	302·6	144·6	47·8	97·9	32·3	60·1	19·9
1966	284·6	131·6	46·2	93·1	32·7	59·9	21·0

% change in
Nos. at work
1957–65   −2·7%   −13·3%   —   nil   —   +30·0%   —

Notes: 1. Nos. at work exclude persons on the payroll of companies absent due to holidays and other causes. Two part-time female workers are counted as one unit.
2. All figures are for the end of the first week in December each year.
3. Sub-totals do not always add to overall totals due to rounding.

Source: Iron and Steel *Annual Statistics* (Iron and Steel Board/BISF).

the industry's employment is highly susceptible to changes in the level of industrial activity, it is probably unwise to read too much into the apparent decline in numbers at work between 1957 and 1966; for

1957 was a year of high activity in the industry whereas 1966 was one of low capacity operation. A better, but not perfect guide to the longer run trend can be obtained by comparing 1957 and 1965, both of which were years of high capacity utilization—though of course capacity had increased considerably in the interim. On this basis the real employment decline for the period can be estimated at around 2 to 3 per cent.

The most striking change between 1957-65 was the increase of 30 per cent in the numbers at work in the administrative, technical and clerical grades, resulting in a rise of their share in total employment from 15 to 20 per cent. Among process workers on the other hand, there was a fall of about 22,000, or just over 13 per cent and a decline in their relative share of total employment from 53·6 to 47·8 per cent. The general and maintenance category remained very stable both absolutely and in relation to total employment.

More detailed information about the decline in process workers' employment is given in Table 5.7, which shows the changes in numbers at work in the various process and product stages between

TABLE 5.7

Process Workers, by Process Grouping: 1957 *and* 1965[1]

| Process Group | 1957 | | 1965 | | Percentage change |
	Nos. at Work	As % of total	Nos. at Work	As % of total	in Nos. at Work, 1957–65
1. Iron-ore Mines and Quarries	4,610	2·8	2,180	1·5	−52·7
2. Coke Ovens at Blast Furnaces	5,230	3·1	4,910	3·4	− 6·1
3. Blast Furnaces and Sinter Plants	13,230	7·9	10,370	7·2	−21·6
4. Steel-melting Furnaces, etc.	23,690	14·2	20,280	14·0	−14·4
5. Rolling Mills	49,720	29·8	49,370	34·1	− 0·7
6. Sheet Mills	16,110	9·7	12,650	8·7	−21·5
7. Tin-plate manufacture	5,030	3·0	3,130	2·2	−37·8
8. Forges and ancillary processes	6,740	4·0	4,820	3·3	−28·5
9. Steel Foundries	20,900	12·5	17,470	12·1	−16·4
10. Wrought Iron manufacture	1,030	0·6	180	0·1	−82·5
11. Steel Tubes, Pipes, etc.	20,440	12·3	19,230	13·3	− 5·9
Total	166,730	100·0	144,590	100·0	−13·3

Note: 1. Figures are for the first full week in December of each year.
Source: Iron and Steel *Annual Statistics* (Iron and Steel Board/BISF).

1957 and 1965. Line 1 of the Table shows a large proportional reduction in the numbers at work in iron-ore extraction. This is largely due to the consolidation of mining into two main areas of the country and an increase in mechanization. Lines 2 and 3 (coke ovens, sinter plants and blast furnaces) again show significant proportional reductions, and while greater output of coke and iron-ore has been achieved, technical improvements and increased mechanization in these processes have permitted some reduction of manpower. Technological change is again a factor in the explanation of the reduction in numbers employed in steel-melting (line 4). Employment at the rolling stage (line 5) has been least affected, though there have been very considerable changes in the process resulting in higher outputs and especially higher quality products. The sheet mills (line 6) have again experienced an expansion of capacity and output, but a high proportion of the new capacity has come from the wide strip mills which are less labour-intensive than the older mills. Tin-plate manufacture (line 7) was subject to a massive plan for rationalization which ended about 1959,[1] since when employment has remained fairly stable despite a substantial increase in output. However, relatively little will be said about this in the following chapters, or about the groups covered in lines 8 to 10 of the Table, which deal with the shaping of finished iron and steel by means other than rolling. The production of steel pipes and tubes has again benefited from developments in the rolling of narrow strip which have been accompanied by manpower savings per ton of output.

Overall, the picture is one of increasing output at virtually all stages of production over the period, which has been achieved by the replacement of obsolescent plant and the introduction of new plant incorporating modern techniques and greater capacity. The rise in output has, however, not been sufficient to maintain employment for process workers in view of the labour-saving character of many of the changes and despite a reduction in the standard length of the working week, though it is obvious that there has been some compensatory movement in the marked increase of employment in administrative, technical and clerical grades.

One final point is that steel is mainly an employer of male labour. Female employment has been fairly constant at about 18,000 to 19,000 in the past few years, though there has been a slight suggestion of an increase (and hence an increase in the share of female

---

[1] For a historical account of the tin-plate industry and its rationalization see W. E. Minchinton, *The British Tinplate Industry: A History*, Oxford University Press, Oxford, 1957.

employment in the total). About two-thirds of female employees are in the administrative, technical and clerical grades.

VI. INDUSTRIAL RELATIONS

The following brief résumé of industrial relations and collective bargaining arrangements in the iron and steel industry covers the following subjects: (i) trade union organization; (ii) employers' associations; (iii) the main principles of collective bargaining; (iv) arrangements relating to the manning of plant and seniority rules; and (v) disputes procedures. The description, of course, relates to the pre-nationalization period.

(i) *Trade Union Organization*[1]

The main distinction to be drawn is that between craft workers and process or production workers. The principal trade union, with a membership of about 120,000, is the Iron and Steel Trades Confederation (known generally in the industry as the Confederation), which organizes production and ancillary workers in all sections of the industry, and even some clerical and supervisory grades.[2] But its coverage of production workers is far from complete. Although in Scotland workers employed at blast furnaces are organized by the Confederation, in England and Wales they are members of the National Union of Blast Furnacemen, Ore Miners and Kindred Trades (NUB). In some places craft workers belong to the Confederation, but in general these workers belong to the appropriate craft union such as the Boilermakers' Society, the Amalgamated Engineering Union, the Electrical Trades Union, the Plumbing Trades Union, etc. For collective bargaining purposes, ten of these craft unions joined together in the National Craftsmen's Co-ordinating Committee, to negotiate with the main employers' association,

[1] A number of changes have occurred in the organization of trade unions and employers' associations since this section was first written. Since this described the situation at the time of most of the case studies, no changes have been made, but the following points should be noted:

    (a) the NUGMW is now known as the General and Municipal Workers Union;
    (b) as a result of a merger with foundry workers the AEU is now the Amalgamated Union of Engineering and Foundry Workers (AEF);
    (c) the Iron and Steel Trades Employers' Association was disbanded in March 1968 following nationalization of the major part of the industry.

[2] Where this occurs, the grades affected are organized in separate branches, subject to independent negotiation.

the Iron and Steel Trades Employers' Association. One prominent craft union, however, remained outside this group, namely the Amalgamated Union of Building Trade Workers (AUBTW) which negotiates separately for its members who are primarily brick-layers.

In addition, there is an important representation in the industry of the two large general unions, the Transport and General Workers Union (TGWU) and the National Union of General and Municipal Workers (NUGMW). While these mainly concern themselves with the organization of general labourers and workers providing services, such as transport, they do have a sizeable membership in parts of the production process, such as the sheet trade.

The trade union situation in the industry is obviously complex, with eighteen unions being represented. A high proportion of the labour force is unionized, and while in the craft and general unions the normal form of branch organization exists, practice within the Confederation is rather different, with branches usually covering members in a department or engaged on a single process or class of work.

(ii) *Employers' Associations*
The main employers' association was the Iron and Steel Trades Employers' Association (ISTEA), which dealt with the heavy steel trade and grew out of a large number of associations based on product or area lines. The Association was divided into four geographical Divisions, each of which was almost entirely autonomous, having its own Council which was free to enter into agreements subject to the proviso that they did not thereby contravene national agreements or produce serious repercussions in other Divisions. There remained a number of small employers' associations, organized along product or area lines, which negotiated for their members on wages and conditions. The employers' associations all belonged to the Central Council of Iron and Steel Trades Employers' Associations, which acted as a forum for the discussion of national issues and provided the safeguard that a member association would not conclude an agreement without regard to its repercussions elsewhere in the industry. Most, but not quite all, of the main companies in the industry were affiliated to at least one Association.

(iii) *Collective Bargaining*[1]
It is clear from the organization of trade unions and employers'

[1] Again, this must be regarded as an outline of the position prior to nationalization.

associations that the steel-industry collective bargaining structure, prior to nationalization, was complex. To some extent the complications were reduced by the grouping of unions, as in the National Craftsmen's Co-ordinating Committee. But still, ISTEA itself had to negotiate at national level with five separate unions or union groups, while other associations had to negotiate with at least two, and often more, unions.

(a) *Production workers.* Most production workers had four elements in their pay. The base rate, firstly, was negotiated primarily at works level, after consultation with the appropriate Association. (This base rate used to be subject to a sliding scale addition, introduced early in the century to allow workers to share in the benefits arising from favourable trading conditions. However, with the advent of price controls the sliding scale addition was incorporated into the base rate.) Secondly, there was the cost of living payment, tied to the Ministry of Labour's Index of Retail Prices. Adjustment in the gearing was negotiated at Association level. The third element was the tonnage bonus, normally negotiated at works level, except for melters.[1] Finally, there were the 'extras' such as shift work premia, overtime rates, week-end working payments, abnormal conditions rates and so on. These were mainly matters for negotiation at national level.

As a general guide, it can be stated that matters of more general application, such as hours of work, paid holidays and working conditions, were negotiated by the Association and appropriate union officers, at national or divisional level. Matters of less general applicability were passed to works level, where they were negotiated by the company and the appropriate union branch, but some consultation often took place at Association level. Thus a considerable autonomy was left to the works themselves.

(b) *Craftsmen.* Again, generalization is impossible. Workers coming under the National Craftsmen's Co-ordinating Committee were subject to a national agreement covering basic wages, overtime premia, holidays with pay and guaranteed employment, but they would usually also participate in a tonnage bonus system of pay-

---

[1] Until 1965 a most important feature of the bargaining system was the 1928 Melting Rates Agreement between the Confederation and ISTEA. This agreement covered specified workers in open-hearth melting shops on a nation-wide basis, and comprised mainly a straight tonnage payment geared to the productive capacity of the open-hearth furnaces, with a small base or datal payment. The Agreement lapsed, however, in 1965 and negotiation on melting rates was passed back to works level.

ment negotiated by district committees. In other cases where the local employers' associations did not negotiate with the craft unions, the normal district rate for each type of craftsmen would usually be paid. Payments related to cost of living index changes and to abnormal working conditions were also common. Shift working of various types was common among maintenance craftsmen, and enhanced earnings here derived not from differences in rates but from paying for added time known in the industry as 'gift hours' for afternoon and night shifts, and a 'rota working extra' bonus.

(c) *Labourers*. Labourers directly associated with production jobs were normally treated as production workers. General labourers, on the other hand, seldom received a tonnage bonus, and their pay was made up of a basic or minimum rate and a cost of living bonus, both negotiated at national level, plus payment for overtime or other extras.

The fact that national agreements, where they existed, were concerned with broad matters, and the fact that much scope for negotiation was left to negotiators at plant or district level, meant that as a rule each case of technological change was dealt with as and when it occurred at plant level, and while experience elsewhere might be sought out, there was no necessary establishment of precedent which obliged a firm to follow some other firm's practice. The number of men employed on new equipment, and the structure or level of payment, could vary from case to case. Rates of pay for similar jobs within the industry could differ significantly from region to region and further differences might be due to differences in plant conditions or in ancillary plant.

(iv) *Manning Agreements and Seniority Rules*
The manning of particular items of plant and the rates of payment for operatives were delegated almost entirely to negotiation at works level. Furthermore, once manning and rating had been agreed, it was a convention that unless a 'change of practice' made it necessary, that agreement would stand. Following a change of practice, new manning, revised rates or a combination of these could be discussed. The now defunct 1928 Melting Rates Agreement was an exception to the general principle, for although the manning varied according to the size of the open-hearth furnace the *ratio* of payments for each grade of operative was laid down on a national basis, the actual sum to be shared depending on the output of the furnace. A rather similar agreement, negotiated by ISTEA and the Confederation,

131

covered the manning and payment to operators on the LD steel-making process.

Another aspect of the manning problem following technological change arises from the multiplicity of unions, and the lack of any standardization in their coverage within the industry. The operating rule over manning has been that production jobs will be manned by members of the Confederation and the NUB and in some cases also by members of the TGWU and NUGMW. But with certain kinds of technological change, the definition of a production job has been, at best, difficult, and there have been important disputes between the Confederation on the one hand, and unions such as the AEU and the NUB, which are considered later.

Though the trade unions in the industry do not control the intake of labour as closely as in the printing industry, there is nevertheless a very important form of promotion control practised by the Confederation. Each works, and indeed most production departments, have quite distinct work groups, specialized in function or location. Newcomers to the works (normally whether or not they have been previously employed in the industry) will enter the production processes at the grade of labourer; though this does not apply to craftsmen and other workers outside the production grades. As they acquire seniority and as vacancies open up on the appropriate promotion ladder, labourers will move on to the bottom rung. Thus promotion to increasingly skilled jobs on the production side is governed strictly by the seniority principle, and while management normally reserves some right to prevent automatic promotion of this kind on grounds of lack of ability or lack of qualifications, the seniority rule dominates the scene. This, as we shall see subsequently, has important implications for the implementation of technological change.

(v) *Disputes Procedures*

The considerable scope for works negotiation gave rise to a unique and fairly successful method of dealing with disputes. Every effort was made to settle issues at works level, through negotiation between management and branch representatives. If this failed the Union Divisional Officer could be called in, and he would consult first with management, and secondly, if necessary, with his Head Office and the Employers' Association concerned. Beyond this stage, there was an established procedure for joint negotiation. A typical arrangement was that of the Heavy Steel Trade, which involved a Neutral Committee (constituted from two workers' representatives and two employers' representatives, all from works not associated with the dispute). Normally this produced a decision, but when it did not, reference was made to a National Joint Sub-Committee, and finally resort could be had to arbitration by a single assessor or an independent chairman with two assessors, whose decision was final. The use made of the latter was very infrequent, and official stoppages have been few.

VII. SUMMARY

From this discussion, a number of characteristics of the industry emerge which are likely to have a bearing on its experience of technological change and on its ability to cope with the problems of change.

(i) The industry is subject to influence from international competition and because of this it is important that it should keep abreast of current technological developments. The home industry has contributed to these developments itself but has also benefited from new technological developments originated abroad.

(ii) Iron and steel is a heavy industry, basic to the economy and sensitive to changes in the level of business activity: output and employment fluctuate a good deal in the short run, frequently leading to low capacity utilization. Yet because of the industry's basic role in the economy it has been considered essential that capacity should be adequate to meet peak demands. This, together with the interdependence of the various stages of production and the long run growth of demand, has tended to result in bunched investment programmes such as that of the early 1960s.

(iii) The industry comprises a number of stages of production ranging from the extraction of iron-ore to operations which are closely akin to general engineering, and it therefore embraces a number of quite different branches of technology, each of which has experienced some degree of technological advance, though progress in some areas has been much more extensive than in others. The research and development organizations of the companies, as well as the industry's central research association, have in recent years been spending increasing amounts in the effort to further this progress.

(iv) The manual labour force in the industry is divided into two groups, the production workers and the general and maintenance workers. This division is accentuated by the fact that production workers are generally (though not completely) organized by the Confederation and the NUB, while the remainder are represented by a large number of unions. While the labour force as a whole has been declining very slowly, the production workers have borne the brunt of the fall, and the proportion of general and maintenance workers has remained quite stable. The white-collar workers in the administrative, technical and clerical group have, on the other hand, been growing in numbers and in their share of total employment.

CHAPTER 6

THE TREND OF TECHNOLOGICAL CHANGE

I. INTRODUCTION

The aim of this chapter is to identify and analyse the main changes in technology that have affected the iron and steel industry in recent years. Inevitably, a host of minor developments have had to be excluded, as have changes in technology which, though important in their own right, are limited to small sectors of the industry. As mentioned earlier, steel production involves three main stages, and the principal object here is to indicate the character of changes at each of these stages. Attention is also given to the broad employment effects of change at each stage, but more detailed consideration of the adjustment problems encountered is held over to Chapter 7.

Before proceeding to the individual stages, a brief overview of the process may be useful. Steel is made from iron by a refining process which reduces the carbon content and removes other impurities.[1] The iron itself has to be produced from iron-ore, and the production of iron, usually by blast furnace techniques, is the first of the three stages. The blast furnace produces molten iron which is transferred to the steel-making stage, where a variety of techniques are used for refining. Conventional practice requires the molten steel produced in the second stage to be partly cooled in ingot form, the ingots then being transferred to the roughing mills, where they are rolled into slabs or billets. These are usually subjected to further rolling and finishing operations before being prepared for inspection and dispatch. Historically, the separation between production stages has been marked and has involved the industry in heavy capital and running costs connected with the advance of the product from stage to stage: for example, cranes, moulds, and heated vessels have been needed to retain the hot metal and keep it at high uniform temperatures to facilitate the next set of operations. As we shall see, much of the technological development in recent years has

[1] Iron can be generally regarded as having a carbon content of over 2 per cent, while steel has less than 2 per cent carbon.

been concerned with the reduction of these discontinuities in production.

The remainder of this chapter takes the following form. Section II considers the changes in technology that have been introduced in the production of iron. Section III reviews the main techniques of steel-making and outlines the trends in their utilization, showing a marked change in the relative importance of the different processes. Section IV is concerned with the evolving technology in rolling and finishing, and in Section V a short account is given of the effect of these changes on employment.

II. PRODUCING IRON FOR STEEL-MAKING

Iron for steel-making in this country is almost entirely produced in blast furnaces. The furnace comprises a tall shaft, which is filled with lumps of iron-ore, limestone and coke. The coke near the bottom is ignited by a blast of hot air. As the hot gases pass up the shaft, heat is transferred to the rest of the burden and the resultant chemical processes produce a fairly pure form of iron, plus slag and combustible gases. The iron is drawn off and is usually poured into ladles for transfer to the steel melting shops.

The basic capital equipment comprises coke ovens, hot blast stoves, ore preparation plant and blast furnaces. The main changes in technology relate to the operation of the blast furnaces and burden preparation. First, furnace operation has become more mechanized, especially in the charging of bulky loads: increasing use has been made of automatic or semi-automatic control and weighing of the materials being charged, with a corresponding increase in instrumentation, and a reduction in purely manual labour. Secondly, furnace practice has changed due to the use of additional fuels, notably fuel oil, powdered coal and oxygen enriched blasts. Until 1961 the use of oxygen for blast enrichment was negligible (about 0·25 million cu. ft. per annum) but by 1963 consumption had risen to 4·8 million cu. ft., though subsequently falling back a little. This suggests a very high rate of technical diffusion, coupled with a period of intensive experimentation, with a final equilibrium level becoming established rather below the peak. Thirdly, there are the improvements in burden preparation. Efficient blast furnace operation requires the exclusion of fine materials which tend to choke the process. Traditionally, coke has been used to ensure adequate permeability, but because of its relative cost, efforts have been made to reduce coke input per ton of iron output.

Because of the permeability factor, any reduction in coke content must be balanced by some other change, and widespread use has been made of sintering for this purpose. Sinter is an agglomeration of fine ores to form uniform 'lumps' of iron-ore of a sufficient size to give good permeability: the sintering process is fairly capital intensive but the capital expenditure is not great relative to other plant investments in the industry, and it is usually within the works. Some indication of the development of sintering is given by the fact that in 1955 there were 22 sinter strands in use, compared with 46 at the end of 1965, while sinter production over the same period rose from 7·8 to 21·3 million tons.[1]

As a result of these changes, the consumption of coke in blast-furnaces, per ton of iron produced, fell from 19 cwt. in 1957 to 13·6 cwt. in 1965, with fuel injection rising from virtually nil to 0·3 cwt. over the same period.[2] Furnace productivity increased substantially at the same time, the average annual output of iron per furnace rising from 145,000 tons in 1957 to 263,000 tons in 1965, though in part this was attributable to the fact that larger furnaces were being built while some older, smaller furnaces were being withdrawn, as well as to changes in practice. These developments enabled existing furnaces to achieve considerably higher productivity at relatively little extra capital cost since the amount of plant modification required was slight. But higher furnace output would only be worth while if demand was increasing or if it allowed existing plant of an older vintage and with higher operating costs to be closed down, including ancillary plant. In the event, between 1957 and 1965 there appear to have been 63 blast furnace closures.[3] As might be expected, most of these occur in years such as 1959 and 1962–3 when the industry was relatively depressed; when demand is at high levels use will be made of all existing capacity, while during slacker periods of activity the opportunity will be taken to remove obsolescent or high cost plant from commission.

Between 1957 and 1965 demand for iron was growing at about 2 per cent annually, but in addition there was a recognized need for

[1] Iron and Steel *Annual Statistics* (Iron and Steel Board/BISF). An alternative to sintering that is growing in importance is the use of pelletized ore, performing much the same function: for fuller discussion, see Iron and Steel Board *Special Report* (*Development*), 1964, pp. 117–18.
[2] Iron and Steel Board, Annual *Reports*.
[3] From figures supplied by the BISF. Some caution is due in using these figures since it is often difficult to distinguish true closures from those coupled with rebuilding. So far as can be determined, the quoted figures relate to closures where no replacement occurred.

expansion to meet longer-term growth, so that there was adequate incentive for producers to take advantage of the technical advances. To some extent this was achieved by the construction of new furnaces, especially in the new steelworks like Spencer Works which were brought into commission during the period. There was also a gain in capacity by the straight replacement of older, smaller-scale furnaces by larger scale plant subject to major economies of scale.[1] But probably the main factor in quantitative terms was the improved productivity of existing plant, following relatively minor modifications and the use of improved burden preparation. The net result was that over the period 1957–65 the number of furnaces *in use* declined from 98 to 66, while average annual furnace output rose from 145,000 tons to 263,000 tons.

III. STEEL-MAKING

Steel-making can be undertaken in a number of ways which depend on quite different technological principles. In each case the common element is the production of steel from iron or scrap or some mixture of these by a process of refining involving the heating of the charge to release the unwanted impurities. The available processes can be conveniently discussed under three main headings: open hearth, electric, and converter processes.

(i) *Open-hearth Processes*

Open-hearth steel-making became the dominant process in this country in the late nineteenth century, and even in 1965 open-hearth furnaces accounted for over 60 per cent of crude steel output. The open-hearth plant is a large chamber lined with refractory brick, at the bottom of which lies a shallow hearth to receive the charge. Burners on the walls of the chamber direct a flame on to the charge which melts at about 1500–1600°C, releasing impurities that are removed by oxidation or in the slag that forms. Iron-ore and/or mill scale are also added. The open-hearth process has the advantage of being able to use cold (or partly heated) iron, can use a wider range of grades of iron than the earlier converter processes and perhaps most importantly, it can utilize cold steel scrap in quantity.

One apparent disadvantage is the long cycle time of production, an average tap-to-tap time being about eight to ten hours. However,

[1] For a discussion of scale economies in iron-making, see C. Pratten and R. M. Dean, *The Economies of Large-Scale Production in British Industry*, Cambridge University Press, Cambridge, 1965.

over the years the size of open-hearth furnaces has increased considerably, and in the 1950s installations with capacities of over 500 tons were made, though the largest in this country is 400 tons. The large furnace size and the lengthy production cycle mean that large batches of ingot steel are produced periodically. This periodicity is overcome by the operation of a number of furnaces phased to produce a regular flow of ingot steel to the subsequent rolling stages, though this does mean some loss of the scale economies available on larger furnaces.[1] Also, the large batches of steel which the very largest furnaces could produce would require additional rolling capacity which, in the circumstances of the relatively small British market and its large number of competing works, would not be economically justified. The basic problem here is that ingots going forward for rolling have to be maintained at a high and uniform temperature, and this is an expensive operation, so that most producers have settled for a larger number of furnaces of lower capacity than is technically possible, to provide greater co-ordination between production stages. In any case, the slow speed of the process has traditionally had the advantage of permitting careful temperature control and a consequently high degree of accuracy in metallurgical terms.

Recent improvements in open-hearth practice have included oxygen enrichment to accelerate melting and the refining process, reducing cycle time and increasing furnace productivity. The Ajax furnace was pioneered by the United Steel Companies to reduce problems of spilling molten metal and slag during blowing, and increases in productivity of almost 100 per cent were achieved. Although these innovations have improved the productivity of plant already built—especially furnaces of recent vintage with relatively high capacities—the improvements are generally not regarded as competitive with the new oxygen-converter processes, and it is unlikely that any new open-hearth furnaces will be built. Nevertheless, these changes in practice have had the effect of prolonging the economic life of some open-hearth furnaces and thus delaying the introduction of oxygen converters to some extent.

(ii) *Electric Steel-making.*
There are two types of electric steel-making: the induction process and electric arc refining. The first is used only on a small scale for some types of specialized steel, and need not concern us here. The

[1] For some indication of the available economies of scale, see Pratten and Dean, op. cit.

electric arc process is, however, of major importance. Although the principles of electric steel-making have been known for a long time, it was only after the costs of electric power had been greatly cut, relative to those of traditional energy sources, that the process became commercially important.[1] But then innovations and improvements in the design and performance of such furnaces made possible a reduction in electric power consumption per ton of steel. In the earlier stages of development, electric arc furnaces were used mainly for the production of special steels and alloys. More recently, increasing use has been made of electric arc furnaces for the production of common steels, and the size of the installation has increased rapidly from around 20 to 30 tons up to well over 100 tons.

The electric arc furnace involves many of the features of the open-hearth furnace, except that it is normally circular rather than rectangular in shape. Heat is supplied from electrodes which are lowered until contact is made with the charge, which is normally composed entirely of steel scrap though there has been experimentation more recently with a charge of hot metal. Oxygen injection is frequently practised to aid in the acceleration of carbon removal and in the recovery of valuable alloying materials where alloy steel is being produced. The electric process allows very close temperature control, favouring the production of high quality and alloy steels, but more common steel is now being produced in this way than formerly. The cycle time is considerably less than the open-hearth method, but still much higher than in the oxygen-converter processes.

(iii) *Converter Processes*
The characteristic feature of converter methods of steel-making is that air or gas is blown through a converter vessel containing molten pig-iron, raising the temperature and allowing the refining process to take place. This is a different technique from open-hearth and electric arc steel-making where *external* sources of heat are applied. The Bessemer converter was the first in the field, beginning around 1856, but it never acquired the dominance over open-hearth in Britain that it achieved in Western Europe. The *Bessemer converter* was small (and remained so) compared with open-hearth furnaces. Most converters had a capacity of about 25 tons, but a few were built of 50 tons and, in other countries, vessels as large as 70 tons have been used. Although the capacity was

[1] Cf. *Comparisons of Steelmaking Processes*, pp. 10–11, United Nations (ECE), New York, 1962.

relatively small, the cycle time was short, with a blowing time of less than half an hour for a 25-ton vessel and a proportionately short tap-to-tap period. Despite a number of innovations, including the use of an oxygen-enriched air blast or an oxygen-steam combination, the Bessemer process is now virtually unused in this country, primarily because the process is almost entirely dependent on molten iron and is able to use only a very small quantity of steel scrap, which acts as a cooling agent.

However, the essential feature of the Bessemer process, the blowing of air or gas through molten iron, has been adopted in a number of newer converter processes developed in the post-war period. Of these two are of importance here: the Kaldo process, developed in Sweden, and the LD process, invented in Austria. Both have been used in this country but LD has quickly gained an advantage and is likely to be the main steel-making technique here for some time to come.

The *LD process* was the result of research into the needs of Austrian steel-makers to find a technique which could make use of locally available iron-ore with a low phosphorous content.[1] It was initially restricted to low phosphorous iron and to the production of common carbon steels, but subsequent developments have greatly improved its flexibility. The LD technique involves the injection of technically pure oxygen into a vessel containing molten iron. The vessel remains stationary during the blow (which distinguishes it from the Kaldo process) and the oxygen is blown in from the top of the vessel (which distinguishes it from the Bessemer converter which had air blown in from the bottom). The blowing time is short, averaging about 40 to 50 min., and the process can use up to 35 per cent[2] steel scrap which improves its economic performance. The development of the process has been rapid. The first laboratory LD vessel had a capacity of 3 tons, and the first pilot plant a maximum capacity of 15 tons. Ten years later, vessel sizes of 100 tons were common and much bigger vessels still were operating or in course of con-

[1] The diffusion of the LD process has been the subject of some academic discussion. Cf. Walter Adams and Joel B. Dirlam, 'Big Steel, Invention and Innovation', *Quarterly Journal of Economics*, May 1966; and the article by the same authors, 'Steel Imports and Vertical Oligopoly Power', *American Economic Review*, September 1964. Comments on the latter were published in the *American Economic Review*, March 1966. The international diffusion aspects were taken up by G. S. Maddala and P. T. Knight in 'International Diffusion of Technical Change', *Economic Journal*, September 1967.
[2] Recent experiments suggest that the scrap proportion can be raised as high as 50 per cent.

struction, especially in countries where large markets were available.[1]

The *Kaldo* process also involves a blowing of oxygen on to and through the surface of molten iron, but in this case the converter rotates at up to 30 revolutions per minute, allowing better contact between slag and metal and more precise control of conditions inside the vessel. It can use a wide range of pig-iron grades and can absorb relatively large quantities of scrap (up to 45 per cent) and iron-ore. Unlike most steel-making processes, the pilot plant for the Kaldo process was on a large scale, with a capacity of 25 to 30 tons, and within a very short time period new installations of well over 100 tons were being made. The time taken to refine the charge is less than one hour, and the resultant low cycle time allows a fairly continuous flow of steel to the finishing stages. A wide range of qualities of steel is possible, since it is amenable to close instrumental control. But for a variety of reasons the Kaldo process has been used in few British works and technical difficulties or product requirements have caused at least one firm to convert a Kaldo vessel to LD.

(iv) *Other Developments*

Brief mention should also be made of other developments. Possibly of most significance in the long run is *spray steel-making*, in which a falling stream of molten iron is 'atomized' by oxygen jets and instantaneously refined. Laboratory experiments led to the solving of several major problems, and a large-scale trial was successfully completed. Other trials are now in hand in some larger works but are encountering difficulty. The potential of this process seems considerable since the possibility now seems to be present that molten iron can be run directly from the blast furnace to a steel-making process and from there to continuous casting plant.[2] This would introduce a much greater continuity into the industry's production process, leading through, virtually without a break, from the production of iron to the manufacture of semi-finished products.

[1] The Steel Company of Wales, in 1965, received permission to build three 270-ton LD vessels. though the construction of one of these was subsequently postponed.

[2] Continuous casting is discussed in Section IV below. In fact just such a development scheme, leading to a fully continuous works system, was put forward by the Milton Hematite Ore and Iron Company in 1966. Under this proposal a full-scale spray steel-making plant and a continuous casting plant for billet production would have been built. But the scheme was turned down by the Iron and Steel Board as being inconsistent with the industry's need for concentration of production in fewer, well-located works. For a more detailed statement of the argument, see 1966 *Report* of the Iron and Steel Board, para. 115.

Furthermore, the capital expenditure would probably be small in relation to the cost of LD and electric furnaces.[1] At present, however, both technical and economic questions remain to be answered.

Scope may also exist for developing the *Fuel-Oxygen-Scrap* process, developed (like spray steel-making) largely under the auspices of BISRA. This process promises to cut the cost of producing steel from cold metal, and a fairly large pilot furnace of 80 tons capacity has been built to determine its commercial viability. A brief reference is due to *vacuum degassing*, a process applied at the end of the steel-making stage to produce 'cleaner' steels. The application of a vacuum at this point is costly and is justified only for high-priced special steels: but since this sector of the industry is rapidly growing the importance of the process seems likely to increase.

Additionally, a continuous stream of improved practices and developments in equipment is to be observed in the technical journals of the industry, the Reports of BISRA and elsewhere. The main range of techniques currently in use and likely to be relevant for some time to come have now been covered. We must now see something of the pattern of their utilization in Britain.

(v) *Trends in Utilization*

Although open-hearth furnaces still occupy a dominant role in steel production the picture has been changing quite rapidly, and most of the new investment in steel-making plant has been directed towards oxygen converter processes and electric arc furnaces. There are two ways in which the change can be measured. First, we can observe the changing importance of the proportion of total ingot steel output coming from each of the main process types. And secondly, we can examine changes in the number of furnaces or vessels of each type in existence. These trends are shown in Tables 6·1 and 6·2.

Table 6.1 shows that the share of *open-hearth* furnaces in total ingot output fell from over 92 per cent in 1938 to less than 60 per cent in 1966. One must remember, however, that especially since 1938 (when ingot output was about 10 million tons) and even since 1961, steel output has been increasing, so that tonnage output from open-hearth plant has risen considerably since 1938 and since 1961 has fallen less rapidly than the percentage share figures suggest. The *Bessemer* process never matched the open-hearth in importance during the period in question. Although exact figures are not

[1] It has been estimated that capital costs of spray steel-making might be about one-half those of the LD process, for similar annual capacity.

TABLE 6.1

*Contribution of Various Processes to Total Ingot Steel Production
for 1938 and 1961–6*

Percentages

Type	Open-hearth		Converter			Electric		Total[1]
Year	Acid	Basic	Kaldo	LD[2]	Other[3]	Arc	Induction	
1938	16·4	76·2	—	—	5·9	1·6		100
1961	2·7	82·6	—	—	9·1[4]	5·2	0·4	100
1962	2·1	81·4	—	—	11·0[4]	5·2	0·3	100
1963	1·6	75·9	—	—	14·8[4]	7·4	0·3	100
1964	1·5	70·4	1·0	9·8	7·6	9·4	0·3	100
1965	1·4	63·7	2·0	14·9	6·9	10·9	0·2	100
1966	1·0	59·4	2·5	20·0	5·2	11·8	0·2	100

Notes: 1. Totals do not always add to 100 because of rounding.
2. LD includes LD-AC.
3. 'Other' converter processes are mainly Bessemer in 1939 but from then on include an adaptation of the Bessemer process known as VLN (very low nitrogen) used by the Steel Company of Wales.
4. Figure includes some output from LD.

Source: Estimated from Iron and Steel *Annual Statistics* (Iron and Steel Board/BISF).

available, probably no more than 8 per cent of ingot steel output came from Bessemer type processes in 1960, and the fact that they accounted for as much as 5 per cent output in 1966 is due largely to the existence of the VLN converters which are rather a special case. *Electric steel-making* was used only for special steels in 1938, but contributed about 5 per cent of ingot output in 1961 and nearly 12 per cent in 1966. Of the newer processes, *Kaldo* vessels have only barely made their mark, while the advance of the *LD process* has been spectacular, its share of ingot output rising from less than 1 per cent in 1960 (not shown in the Table) to almost 20 per cent in 1966.

From Table 6.2, striking reductions in the number of open-hearth furnaces since 1955 can be observed. More detailed investigation of the official statistics shows that the shrinkage has been due to the withdrawal of furnaces of low capacity: in fact, as late as 1961–2 the number of open-hearth furnaces of over 300 tons actually increased from 21 to 23. Only two small Bessemer converters remain and this process is now virtually unused, though the Table does not show the continued presence of 4 VLN converters in the Steel Company of Wales (now replaced by large LD converters). The figures for electric arc steel-making are to be explained by the fact that there has been some reduction in the number of small

144

TABLE 6.2

*Number of Furnaces or Vessels in Existence, by Process, for 1955,
and 1961–6 (excluding Tropenas and Stock Converters)*

Type	Open-hearth		Converter			Electric	
Year	Acid	Basic	Kaldo Vessels	LD[1]	Bessemer	Arc	Induction
1955	63	328	—	—	10	161	86
1961	40	306	2[2]	1	10	187	179
1962	35	282	2[2]	6	8	188	206
1963	31	261	2[2]	7	7	188	217
1964	30	234	10[2]	13	7	193	225
1965	26	215	10[2]	15	7	196	232
1966	23	209	9[2]	15	2	194	238

Notes: 1. LD includes LD-AC.
2. Includes 2 Rotor Furnace vessels.

Source: Iron and Steel *Annual Statistics*.

capacity plant and an increase in the number of large furnaces, including several in the range 80 to 110 tons: the growth in the number of induction furnaces is explained by the building of many furnaces of less than 1 ton capacity, used for specialist (alloy steel) purposes. Finally, the growth of the new oxygen processes, especially LD, is marked and further additions are now planned.

Overall, then, there has been a marked change in the pattern of steel production over the last few years. The Bessemer and acid open-hearth techniques have become almost extinct, while the basic open-hearth process is beginning to die back, though its final demise is still some way off, largely due to the improvements which have been made in plant productivity as competition from new processes has grown. The expansion of electric arc furnace capacity partly reflects the growth in demand for special, stainless and alloy steels, but also the fact that common carbon steels can be economically produced where there is a plentiful supply of cheap and good quality scrap. Of the new processes, initially developed abroad, LD has most rapidly gained ground. As had been shown elsewhere,[1] almost 54 per cent of the steel production *increase* between 1956 and 1964 was contributed by oxygen converter processes, while another 31 per cent of the rise came from electric practice. This in itself is an interesting indication of the rate which the new oxygen techniques have been absorbed (especially as the first LD plant, at Ebbw Vale, was not introduced until 1960).

A detailed account of the reasons for this trend in the use of

[1] Maddala and Knight, loc. cit., Table VI.

alternative steel-making techniques is not possible here, but the main factors should be mentioned. Both the oxygen converter and electric arc processes offer very considerable savings in capital cost compared with the open-hearth melting shop.[1] So far as production costs in steel-making itself are concerned, the main cost is materials, accounting for between 60 and 80 per cent of total production costs. Transformation costs (labour, fuel, refractories, moulds, etc.) range from a high of about 25 per cent in electric steel-making—due to the intensive use of relatively expensive electricity—to a low of 11 per cent on LD.[2] An important consideration has been the relative prices of scrap and iron. Scrap has generally been cheaper than iron in the post-war period so that processes able to work on a high scrap ratio have had an immediate cost advantage. The electric arc process can use virtually 100 per cent scrap, open-hearth up to about 50 per cent, LD about 35 per cent (though improvements are still being achieved) and Bessemer very little.

Overall, the choice of techniques since the late 1950s has been between electric arc and oxygen converter processes, with the outcome depending very much on the local availability of scrap, the existence of efficient iron-making capacity and the type of product involved. On this last point, special and alloy steels virtually dictate the use of electric techniques, but while the range of choice for common steels has been wider, there has been a trend here towards closer specification and greater consistency of quality, attributes more readily obtained from processes capable of sophisticated instrumentation, such as electric steel-making and the oxygen converter, and this has again helped to oust the open-hearth method.

IV. ROLLING AND FINISHING

(i) *A Guide to the Processes*

When we turn to the shaping and finishing of steel, the diversity of possibilities and end-products is so great that a straightforward account of each technique is impracticable. Instead, we present an

[1] Kaldo and LD have been estimated to cost about 15 per cent less in capital costs than open-hearth, for given capacity. The electric furnace working with 100 per cent scrap and hence requiring no iron-producing plant, may cost about 25 per cent less than open-hearth, though of course in many cases the iron making capacity may already exist. Cf. Comparison of Steel-making Processes, United Nations (ECE), op. cit., for a full discussion of costs.

[2] ECE estimates (loc. cit.) for a plant of 1·5 m tons annual capacity. It is worth noting that labour costs at the *steel-making stage* are of little importance, accounting for between 10 and 18 per cent of transformation costs or about 3 to 5 per cent of total costs.

outline of rolling and finishing operations and indicate the main changes that have occurred. This approach is the more justifiable because ingot steel, *en route* to its final product destination, usually passes through one or more rolling operations, the basic principles of which are broadly the same.

Although the route of ingot steel through the process may vary, an example of the sequence may be useful. The route is taken up at the point at which the ingot has left the melting shop and lies in the

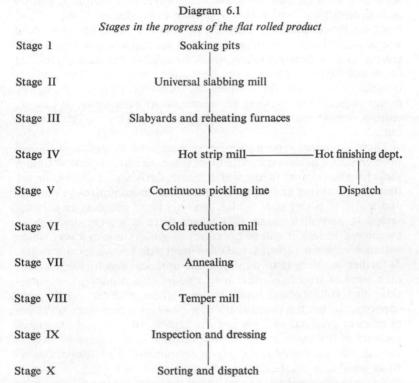

Diagram 6.1

Stages in the progress of the flat rolled product

Stage I	Soaking pits
Stage II	Universal slabbing mill
Stage III	Slabyards and reheating furnaces
Stage IV	Hot strip mill————————Hot finishing dept.
Stage V	Continuous pickling line Dispatch
Stage VI	Cold reduction mill
Stage VII	Annealing
Stage VIII	Temper mill
Stage IX	Inspection and dressing
Stage X	Sorting and dispatch

soaking pit to keep it at a uniform temperature. Diagram 6.1 traces out the possible path from this point. The first stage in the operation is the rolling of the ingot in the *slabbing mill*[1]: slabs are rectangular

[1] Rolling involves the passing of a piece of metal between two rollers, arranged one above the other much as in a large domestic mangle, to reduce the metal to a thickness closer to that specified for the end-product. Generally the thinner the end-product, as with strip and sheet, the greater is the amount of rolling that needs to be done.

147

in cross-section and are the basis of the flat-rolled product such as we are now discussing. Slabs derived from the ingot will then be transferred to the *slab-yards* until ready for further treatment, when they will be reheated and brought forward to the *hot strip mill* for further rolling. After this the product is in a semi-finished form, and may be moved to a hot finishing department in coil or sheet form for dispatch to customers, probably re-rollers. But most steel proceeds to the *pickling process* (an acid treatment to remove scale or rust) and, now cooled under controlled conditions, will be passed through a *cold mill*—to achieve better finish. Cold rolling, however, makes sheet too hard for normal use and it has then to be *annealed* (subject to carefully controlled heating and cooling). In the temper mill, a final rolling is given to achieve the desired surface finish and flatness as well as the mechanical properties. The product is then passed on for inspection, dressing (the rectifying of minor defects) and packing for dispatch. At each stage, of course, cutting, measurement and other operations have to be carried out.

These processes raise a number of economic as well as technical problems. One of the main difficulties is the derivation of an optimum yield for the original ingot, which means devising a programme for the rolling, cutting and finishing of the steel to minimize the number and length of 'short ends' which can only be returned as circulating scrap. It may also mean that where there is a great diversity of customers' orders, it will be necessary to group these in a way which will ensure the most efficient use of an ingot with known specifications. A further problem is to minimize the number of actual operations on a piece of steel to convert it to a designated product. Every pass through a rolling-stand involves time, labour and the use of plant capacity, so that the minimization of passes is a necessary objective of efficient production. This in turn requires that the adjustment of each set of rollers from the passing through of one piece of steel to another should be either easily accomplished (e.g. automatically) or as small as possible.

To these problems have to be added the fundamental discontinuity of the various processing operations, which lead to a need for expensive transfer equipment, reheating furnaces, etc., and the rise in the quality of product required by the industry's customers, expressed especially in a demand for output of *consistently* high quality, which is so important in industries like motor-car manufacture. The industry has responded to these problems in three ways. Firstly, there is the effort to improve the efficiency of each separate

148

stage in the series of rolling and finishing operations, including the better control of reheating, gauge control and the cutting of rolled steel into lengths. This we will define as *stage control*. Secondly, there are the technical advances which have been made in integrating the operations in the rolling and finishing sections of the industry especially in tightening the relationship between order books and the programming of production processes, and in the better co-ordination between individual stages. This we shall refer to as *continuous control*. Thirdly, there is an important area of technological change which involves the much closer physical and *technical integration* of processing stages, notably the development of continuous casting. Each of these is now investigated in rather more detail.

(ii) *Changes in Technology*
At least in the integrated works, but in most other areas of the industry also, the stage of simple mechanization has been achieved and the manual setting and operation of mills is largely extinct. On the other hand, there are some points at which almost completely automatic processing has been introduced, as in the hot strip mill of new vintage Spencer Works. As a general guide to the state of development, it is probably most correct to say that rolling mill technology lies half-way between complete mechanization and complete automation of single processes.

(a) *Stage control.* One highly important area of stage control is in the control of the gap between the rollers through which the steel passes on its way through the mill. Frequent and accurate change of the 'roll-gap' is necessary both because of differences in the required end-product dimensions and because of variations in the size of ingots or slabs entering the mill. As electrical drives for mills were increasingly introduced, the process was speeded up, and manual setting of the controls had to be abandoned. The most recent developments here have been in electronic and hydraulic roll-setting, gauge measurement of the product being continuous, and automatic changes in the roll-gap being made to give a product falling *consistently* within the required margins. There is no need for the operator to take any action during a series of rolls, though it has not often meant a reduction in the mill crew, but has allowed them time for other duties.

The problems become more complex as the frequency of changes in gauge increases, and there has been an increasing trend here toward the use of pre-selecting programmes and more fully automatic on-line computer systems. In the primary rolling and blooming

149

mills,[1] computers have been introduced to select rolling programmes, given the weight of the ingot and the required final size of the product. Similar use is made of computers and punched card systems to operate billet mills, in which changes can be quickly effected in roller speed, roll gap and, in the case of reversing mills, number of passes. Direct computer control of rolling operations has been growing in recent years. This is particularly important in sections of the industry such as wide strip production, where consistently high quality is a major factor for many users. Computer or other forms of automatic control have proved more reliable than manually controlled systems, and in these cases again automatic gauge control and computerized shearing are increasingly common.

Other aspects of automatic control in this sector of the industry include computer control of temperature in soaking pits and reheating furnaces, which not only minimizes power consumption but is becoming capable of selecting and bringing forward steel as it becomes ready for rolling or re-rolling. Some parts of the finishing sectors such as electrolytic tinning and strip galvanizing have been subjected to automatic control, and automatic inspection has been introduced as well.

Despite these changes, the industry still has a long way to go in developing automatic systems of control in finishing, inspection and dispatching, which are still extremely labour intensive. The industry in general recognizes the problems of cost and organization represented by the labour-intensity of finishing, as witness the comment in 1965 Annual Report of the Iron and Steel Board:

'In many steelworks the inspection of both semi-finished products and the rectification of defects still consume a great deal of time and labour; substantial gains in productivity will therefore accrue if these processes can be mechanized or automated.'[2]

Nevertheless, a number of developments have occurred, such as the X-ray inspection unit for billets at the Samuel Fox Works, which automatically locates and marks surface defects. This was viewed as a desirable development for many years at Samuel Fox but the application had to await developments in the electronic field. It is important to notice that this kind of improvement must not only do the inspection job thoroughly and reliably (or orders may suffer) but also rapidly enough to clear the mill. This is a very typical

[1] Just as slabs are the basis of flat-rolled products, blooms are the basis of sections, beams, bars, etc.
[2] loc. cit., p. 55.

instance of the problem faced by steelworks in the finishing departments. It is one of the high costs areas that has been pinpointed, but the technical problems may be resolved only by research and development outside the industry itself.

(b) *Continuous control.* By continuous control in this context is meant the secondary phase of automation technology, in which not merely single processes are automatically controlled, but whole series of processes are interlinked and controlled by one or more computers or other advanced control systems. This need not extend over a whole works—in steel this would be a very advanced step indeed and one which is not yet readily conceivable. It would, however, cover a range of processing operations previously regarded as quite separate from one another. It would involve not only direct control of the product at each stage, but would be capable of 'optimizing' the yield of products by effecting the most economic use of given materials and making any alterations (e.g. in mill-settings, roll-speeds, shearing, etc.) rendered necessary by changing conditions in the processing area.

A good example of developments in this direction is provided by the 'on-line' and 'off-line' computer-control systems in use at Spencer Works (formerly of Richard Thomas and Baldwin) commissioned late in 1962. The main on-line computer operates the hot strip mill and controls and tracks the progress of a slab from the time it enters the slab-reheating furnaces until it reaches the coilers beyond the last of the finishing stands. In terms of our earlier Diagram, it covers the whole of Stages III and IV. The identity and destination of each slab is registered throughout the mill as it enters the system and from there on the whole process is automatically adjusted, to give optimum speed through the mill. The result is a product of consistently high quality, almost certainly unattainable by manual or even semi-automatic operations. In fact the system can be operated manually but only with major losses in speed of operation and quality.

The on-line system is linked to an off-line control system operated by three further computers, which together provide information and control the programme of work. The first computer, covering Stages I and II of our Diagram, and providing linkage back to central administration and forward to Stages IV and VI, provides a check on the feasibility of orders and of spreading peaks in demand: it thus plays a significant part in short run production planning for steel plant and mills. A second computer links Stages II and III with the melting shop, keeping a constant check on the progress of steel

151

through these processes, minimizing wastage and diverting products falling outside specifications to more suitable uses. A third computer is used for operational research and scientific calculations, as well as recording points of production difficulty and contributing to quality control.

This development at Spencer Works is still the exception rather than the rule, and largely reflects the recent origin of the works. It is probably easier to design and bring into operation such a complex system from scratch rather than to superimpose it on a production system already in operation and with its own well-developed procedures. Since a similar completely new works is not likely to be built in this country for some time, further developments of continuous control systems will have to take place in existing works and the problem is likely to be one of finding means of applying continuous control systems as an extension of existing automatic control systems currently limited to single stages of the production process.

In summary, the industry is still well short of fully continuous control in rolling and finishing, and some aspects of such a conception are still beyond the technological horizon. But a great deal has been done in the development of automatic control of separate production stages—themselves often complex, multi-stage operations —and some progress has been made in integrating some individual production stages. In both stage control and continuous control many of the advances have been concerned with improving output quality and consistency and, by making more efficient use of plant time, with cost reduction. Before these kinds of advance can be conceived of at all, a high degree of pure mechanization is essential, and there is no doubt that this sector of the industry has been highly mechanized for some time. This, as we shall see later, has important consequences for employment.

(c) *Technical integration.* 'Technical integration' is the integration of process stages as a consequence of developments in steel plant technology, rather than (as in continuous control) integration by methods which primarily derive from a different technology such as electronics and computers. This distinction should not be taken too rigorously but has relevance in the multi-stage operation of rolling and finishing, where *continuous casting* represents an important example. Continuous casting seems to be on the verge of becoming as significant a development in rolling technology as LD has proved in steel-making.

The initial aim of the process was to convert molten steel direct from the furnace or converter into semi-finished products such as blooms or slabs.[1] It was a substitute for the primary mill stage with the obvious advantage that it did away with much ancillary equipment required in primary rolling. Although a similar process has long been used for non-ferrous metals, the practicability of developing satisfactory plant for casting steel in this way is recent. A number of almost simultaneous developments towards the end of the Second World War led to the establishment in several countries of small continuous-casting plants, mostly working on a pilot scale.

In Britain, the major development work was undertaken for the Government by United Steel Companies, and later by the latter directly. The main emphasis was on continuous casting of billets and for a time the pilot plant was used for commercial production, being replaced in 1960–1 by two new continuous-casting machines. In the late 1950s the need developed for additional bloom capacity and a decision was taken to develop a continuous-casting plant for this purpose: this was commissioned in 1962.[2]

A number of other continuous-casting plants are now in existence in Britain, but most of them are used, as in above example, as supplements to existing mill capacity. The major exception to this was the re-equipment in 1964 of John Summers' Shelton plant (with annual capacity of 350,000 ingot tons), which has no primary rolling capacity, the whole production of semi-finished steel being by continuous casting. This was the first plant in the world to be so equipped on this scale, and marked an important breakthrough, for until then the use of continuous casting had been justified as a means of adding to capacity without going to the vast expense of installing completely new primary rolling mills. But even as this plant was being built, a second generation of continuous-casting machines was being evolved, with less expensive, lower constructions with a curved or even horizontal mould.

Technical problems still persist, especially in cooling and in-

---

[1] Molten steel, more or less directly taken from the converter, is poured from a ladle into a high (180 feet) machine and is drawn down in a water-cooled mould of the required shape.

[2] This plant was built by an engineering company in the United Steel group under licence from a Swiss-registered company, Concast A.G. Although this might seem to be an example of a domestic development being taken up and developed abroad, this is not so. The immediate post-war period saw a number of countries going ahead with continuous-casting experiments, and an international consortium (Concast A.G.) was founded to foster the development. United Steel Companies were a party to this consortium, so were sharing the fruits of some of their own development work.

strumentation, but there is some assurance that continuous casting will gradually replace primary rolling facilities across a wide range of product-types. Indeed, it may well go much further, for current research is investigating the extended application of continuous-casting techniques to the production of some types of more nearly finished products. Success in these areas would eliminate some types of specialized rolling mill, and contribute to a still greater saving on capital installations. But yet this is still too far advanced a possibility for it to be taken into serious account.

V. LABOUR-FORCE EFFECTS

Having now identified the main forms of change in iron and steel technology, we go on to consider their effect on the labour force. The following chapter will deal with some of the more detailed problems of adjustment that have occurred, but here we will be concerned with the broad effects on employment levels and skill structure.

(i) *Iron- and steel-making.*

Relatively little can be said about the effects of technological change on employment in iron-making. Table 5.7 showed a decline of about 3,000 men in blast furnace and sinter plants between 1957 and 1965. (The observed lack of change in coke oven technology is reflected in a relative stability of employment.) Furthermore, we commented upon the growth in the number of sinter plants during the period and while these are capital intensive, there may have been some increase in employment there and a fall of *more* than 3,000 in blast furnace employment proper. Although this is a large numerical drop, representing a decline of one-fifth in blast furnace employment, it was spread out over a period of nine years, but against this we have to set the information about blast furnace closures discussed at the beginning of this chapter,[1] which would indicate a concentration of the employment rundown in a few of these years.

Table 5.7 also showed a fall, over the same period, in the number of process workers in melting shops, but it is far from easy to determine how far this fall of 3,500 (15 per cent) was due to technological change. The year 1957 was, as we have seen, a year of high capacity operation in the steel industry, but by 1958, with the onset of recession, capacity utilization had fallen from 96 to 83 per cent, and the numbers at work had fallen from 23,690 to 20,070, less than the number at work on the same date in 1965. Evidently, changes in

[1] See above, p. 137.

the level of industrial activity are likely to have much more immediate and serious effects on employment than the technological changes which have been occurring. Nevertheless, if, as argued earlier, we can reasonably regard 1957 and 1965 as broadly comparable years of peak activity, the fall of 3,500 in the numbers employed would suggest that technological change has been influential in reducing melting shop employment despite the marked increase in output.

The question is how this reduction has been brought about. The years 1960–2 were the main period of capital investment and even in 1963–4 the amount of investment in steel-making plant remained fairly high.[1] Thus we might expect that, since the period of construction and installation of new steel-making capacity up to the point at which it achieves a regular production routine is lengthy (normally one or two years and sometimes longer), the first labour-force effects of the boosted investment programme would be felt only in 1962–3. In other words, it could be expected that about that time obsolescent capacity would start to be withdrawn as the new plant came into operation; and since the new plant would generally have a higher rated output than the old, there would be a net reduction in the number of plant units in existence. This does seem to have been the case, as Table 6.3 shows.[2]

TABLE 6.3

Closures of Open-hearth Furnaces, 1957–65

Year	1957	1958	1959	1960	1961	1962	1963	1964	1965
No. of closures	6	10	18	25	8	31	25	28	22

Source; Figures supplied by the British Iron and Steel Federation.

Although there is some slight relationship between capacity utilization and the timing of closure here, the main explanation seems to lie elsewhere. The rise in the closure rate in 1958 and 1959 may well reflect the reaction of producers to a decline in demand, taking out of operation plant which had only been kept in existence by high demand. But 1960 seems to have been the year when capacity finally overtook demand, and from then on closures are probably to be explained mainly in terms of the withdrawal of high cost plant as it was replaced by newly commissioned plant. The highest rate of closure came in 1962-5, when we would have expected the new

[1] See Table 5.4, p. 120.
[2] As in the case of blast furnaces these figures have to be treated with some caution.

capacity from the 1960–2 investment boom to be coming into full production. In addition to open-hearth closures, eight Bessemer converters were replaced by LD vessels between 1961 and 1966. Comment has already been made on the withdrawal of small electric arc furnaces.

Thus over the period the withdrawal of old plant was considerable, and though the implicit capacity was not only replaced but increased, employment opportunities undoubtedly declined. It is relevant to observe, however, that even had there been no change in technology over this period—if for example old open-hearth furnaces were simply being replaced by new open-hearth furnaces incorporating minor improvements—employment opportunities would still have declined. For in view of the need for increased capacity, and the known economies of scale in open-hearth practice, replacement would have involved a smaller number of much larger furnaces. In the event, the lower capital and operating costs of LD and electric practice probably had the effect of accelerating the replacement process by making marginal plant uncompetitive, and so techno-logical change undoubtedly had *something* to do with the fall in employment. It is as well that we should be reminded that the new technologies do not constitute the whole explanation.

Further, it happens that the operating team for the major steel-making processes is the same; usually four men with a fifth man perhaps helping out two furnace crews. Because of the fairly constant level of manning even between processes, it follows that changes from one practice to another will not in themselves change melting *team* employment very much, if at all. But there are other very important consequences. First, the new plant units almost invariably have a much higher output than the old, so that, unless demand is increasing especially rapidly, the number of plant units will be less than before. Thus for example 21 open-hearth furnaces at the works of Steel Peech and Tozer were replaced by only 6 electric arc furnaces which raised capacity by about one-fifth. Secondly, the kind of reduction just outlined is multiplied by the fact that the furnace crews form only a small proportion of the melting shop labour force: there is a much greater number of ancillary workers, whose jobs will also be affected by technical change. A change of the kind just mentioned would have the effect of reducing furnace crew employment by about 180, assuming a 21 shift per week system in both cases. In fact the number of jobs which disappeared amounted to just over 900, or five times the pure effect on the melting crews. Thirdly, the number of shifts per week might be reduced following a change, with the possibility of additional shift operation being held in reserve as a longer-term means of adding to capacity: this in itself could reduce employment by one-quarter in the kind of case mentioned.

From one point of view it can be argued that the rising demand for steel has done much to cushion the effects of technological change on melting shop employment. The labour-force reduction averages only 450 per year between 1957 and 1965, which could easily be absorbed without redundancy simply by making use of retirement, let alone normal wastage. But technological change tends to be accompanied by rationalization, involving the closure of small works or whole departments of larger works, so that the incidence of change hits relatively hard in specific places. Partial closures can usually be handled by encouraging transfers to other departments, but the more serious problem is the works closure, where redeployment depends on alternative employment opportunities outside the works, often outside the company itself. There have been several such cases during the period under consideration like that of John Baker and Bessemer in 1964 (discussed in Chapter 7 below).

Although there has been a decline in the number of production workers in the melting shops, there has been some counterweight to this in the expansion of employment in the administrative, technical and clerical grades. The increasing speed of operation in the LD practice requires much quicker sampling and analysis of the charge just prior to tapping, and advanced techniques have been introduced to achieve the necessary information. This is work which requires at least technician grades of skill off the shop-floor, and it is certain that part (though certainly not all) of the increase in technical personnel is *directly* due to the more sophisticated requirements of the newer steel-making processes.

The evidence on simple numerical changes in numbers employed by no means conveys the full extent of the possible problems of technological change. The electric, oxygen converter and open-hearth practices are based on quite different technologies, which raises two difficulties.

In the first place there are major differences in the three main processes, not simply in plant design and layout but in technological knowledge for operators. Each practice, for example, is the subject of separate City and Guilds or equivalent courses. But the main characteristic of open-hearth melters is that they progress from the lowest rung on their particular promotion ladder to the job of first-hand melter almost entirely by seniority, and, at least until recently, formal qualifications were largely unnecessary for accession to the job of first-hand melter. In some ways this is less 'wrong' than might appear. The critical decision in open-hearth practice, the timing of the tap, is that of the chief melter, and since it is based on observation of subtle changes in conditions inside the furnace, coupled with accumulated experience of all kinds of variation, experience and judgement in this case are at least as effective as any formal training. Both electric and oxygen steel-making involve much less reliance on acquired experience, partly because the energy input is subject to greater control, partly because the processes are supplemented by greater all-round instrumentation. (However, promotion is still normally based on seniority.)

Secondly, there is the problem of transfer between processes. The fact that the newer processes require rather less subjective judgement and rely more on observation of monitors and instruments favours the possibility of reasonably rapid retraining. Retraining in these cases involves two separate operations: first, the familiarization of the transferring workers with the physical layout of the plant, its controls and its instruments; and secondly, the instruction of the men in the underlying technological principles. It is the latter which is likely to be the more time-consuming: for example, one firm changing from open-hearth to electric arc practice required its operators to pass a 14-week course in electric steel-making, and this necessity may raise problems when production from existing plant has to be kept up and when shift working is in operation.

In conclusion, there is little doubt that much more emphasis is now being placed on formal qualifications in steel-making practice. The new techniques are both more amenable to formal instruction (coupled with practice) and more in need of properly trained labour. For although some of the 'skill by experience' may be less necessary, the higher speed of operation, the increased need for understanding of instrumentation, and the responsibility for a highly rated, hard-driven plant, are factors which require considerable technical knowledge. Outside the production labour force, there is also some evidence of a rise in the number of maintenance craftsmen associated with the more highly mechanized and instrumented electric and oxygen melting shops. Furthermore, there is also some change in the composition of the maintenance labour force here, with a trend towards more instrument mechanics, electricians, etc., rather than mechanical engineering workers. These are, however, problems to which we shall return in the next chapter.

(ii) *Rolling and Finishing*

In this part of the discussion we turn to a general view of the changes in the rolling and finishing sector of the labour force. Table 5.7 showed that there was virtually no change in the number of *production workers* employed in rolling mills between 1957 and 1965. In the sheet mills there was a fall of about 3,500 and in tin-plate a drop of 2,000, the latter being the effect of the final stage in a major reorganization scheme to eliminate the numerous obsolete, labour-intensive tin-plate manufacturing establishments and to concentrate production in modern facilities operating on a large scale. A fall of about 1,200 was experienced in steel tube and pipe manufacture. In each of these cases the fall in employment was accompanied by an increase in capacity so that production labour requirements per ton of capacity fell considerably.

There are three broad areas of technological advance which are likely to have some association with these employment changes. First, there is straightforward mechanization which, as we have previously suggested, has been going on for some time and has probably little further scope. Nevertheless, mechanization was still being put into effect during the period up to 1965, and since it must be regarded as involving a reduction in labour input per unit of output, it probably has some part to play in the changing employment pattern. Perhaps at least as important as the effect of mechanization on the level of employment is its effect on the structure of employment, for investigation of particular cases suggests that simple mechanization tends to expand the share of semi-skilled employment, at the expense both of skilled employment and of unskilled workers carrying out purely manual tasks.

Secondly, there is the effect of automatic control, which was adopted mainly for reasons of faster operating speed, greater yield per unit of input and improved consistency and quality. The extent to which they affect employment may vary, however, as we can see by distinguishing three types of situations.

(a) *Stage control.* Such innovations are usually added to an established plant to improve productivity in a single stage or sub-stage of production: for example, automatic gauge control and automatic shearing. The servicing and maintenance of the new addition can often be taken on by existing maintenance workers, while the production workers, though perhaps being freed from some duties, may be just as necessary due to faster operating speeds. At most, the jobs of one or two men per shift are all that are likely

159

to be affected, and the change may be one of altered function rather than disappearance of the job itself.

(b) *Complete stage innovations.* These involve the replacement or net addition of plant dealing with a complete stage in production, giving scope for the simultaneous introduction of a number of improvements consistent with 'best practice' standards: for example, the replacement of a complete mill. Here the identification of the pure employment effects is difficult, for the change of process or plant unit itself will often mean changes in the physical scale and output rate of the plant, in plant design and output specifications. The only real comparison in this case is between plants of different vintages and not between the more or less automatic plant as such.

(c) *New works.* The third possibility is the building of a completely new works, such as Spencer Works, or the major rehabilitation of an old works such as Shelton. The constraints on re-equipment imposed by existing plant are then negligible, and a pursuit of best practice methods across the board can be undertaken. There are, however, serious problems when we try to assess the employment changes of such an innovation. The size and structure of the labour force will not vary proportionately with the scale of output, and since the labour content of different product-types varies greatly, a proper comparison would have to be based on works with a similar production-mix. Different works have been found to use different terminology to describe workers' doing the same job, and to adopt different methods of departmental grouping, while there are also variations in the amount of work contracted out. A proper comparison has proved impossible to find, but there is some certainty that the new vintage plant would exhibit a considerably lower standard of manning overall, though the compositional change would not be proportional. The intensive use of automatic control techniques tends to reduce *production* labour input per ton of finished product fairly drastically, but may have the opposite effect on engineering, maintenance and supervisory or managerial personnel. On the evidence of Spencer Works compared with a broadly similar plant of an earlier vintage (though with many modern additions) it seems that the older plant would have rather less than one-quarter of its total labour force as staff, compared with about one-third in the new works. There is also evidence that the internal composition will differ. Among works-grade engineering personnel, a higher proportion will be concerned with electrical maintenance and a lower proportion with mechanical work in the new vintage plant. In addition, works-grade engineering personnel in the newer plant will be roughly equivalent in number to the production work force, compared with an engineering labour force in the older works only one-half the size of the production labour force. The enlarged proportion of staff in the new plant will generally be explained by the greater use of white-collar workers and technicians to operate and service the information handling equipment and on-line automatic devices.[1]

Finally, a brief comment should be made on continuous casting. It has been explained that most continuous-casting plants now in use are supplementary to primary rolling capacity, and are therefore on a relatively small scale and unlikely to have had much in the way of an employment effect. The ultimate labour-force effects could, however, be substantial, if current experiments extend the process to the forming of more nearly completed products, and if the example of Shelton is copied elsewhere. The concentration of a number of hitherto separate processes into a single continuous operation would almost certainly involve a substantial run down in rolling mill employment (including ancillary workers presently engaged in transferring the product between stages). Exactly how big the effect will be, and when it will occur, will depend on the rate of advance of continuous-casting technology and of the optimum plant size, for it is accepted that as the capacity of a continuous-casting plant increases, its labour complement will not increase proportionally.

One of the main conclusions deriving from an examination of these different forms of technological change is that their effects on employment are extremely variable. There is no doubt that in the past, the movement from manual to mechanized operation had considerable effects on employment, though capacity increases often allowed employment in total to remain stable (as in rolling mills) or to decline only a little (as in tin-plate and sheet where mechanization was most important). But it is probable that simple mechanization is more severe in reducing unit labour requirements than the subsequent development of control systems applied at single stages of the production process. These changes are to be regarded less as the substitution of new for old techniques than as partial improvements and although at the margin they may increase capacity, they are less likely to cause existing plant to be withdrawn. If this is so, the current period to 1975 and even beyond may be one in which the effects of technical innovation on employment levels are less severe than those of the earlier period—though at the same time if the

[1] The more integrated the processes become, the more difficult it is to allocate labour—especially staff—to particular departments.

161

growth of capacity is less the positive employment effect will be less strong than before.

There is, however, an additional possibility. Although 'stage' improvements may be more limited in effect than mechanization, it may be that once continuous control over a range of processing operations is achieved, the whole conception of the inter-relationship between process stages is altered, and a further major reduction of process-worker employment may be possible. The evidence to support such a view is still limited but it does seem that the application of partial automatic and semi-automatic controls have had little effect on production workers—though they *have* altered labour force structure and changed job-content; while in so far as one can compare process-worker employment in old and new vintage rolling departments, the dimensions of employment seem to be of a different order. The implication is that piecemeal developments may give less scope for alteration in the labour force than complete replacement or starting from scratch. If that is indeed the case, and if as seems likely no completely new steelworks will be in operation before 1975, the production labour-force effects of technological change in the next few years will be relatively small.[1]

This is not, however, to conclude that there will be no labour-force changes. Even new plant will depart from its original standard manning as operating experience is built up but the scope will be small relative to the possibilities in an older vintage plant with a standard manning geared to plant, processing operations and organizational standards which are no longer relevant to current managerial practice and trade union attitudes in an era of labour shortage, incomes policy and productivity bargaining. The British steel industry, as has been recognized widely for some time, is overmanned by international standards, and not only in the oldest works.[2] Technological change is often used to change standards of manning which could have been altered by other means, and there have been increasing signs, in the economic climate of the late 1960s, that managements in steel (as elsewhere) are making use of these other means. But this is a question to which we return in the next chapter.

In summary, the overall picture is one of declining labour-force requirements per ton of finished output. In part, this has been due

[1] The scope for large scale use of continuous casting, which again could have considerable labour-force effects, would also be limited in this event.
[2] Recent statements by the Chairman of the British Steel Corporation indicate an employment run down of about 50,000 (20 per cent) by 1975: see (e.g.) *Financial Times*, Jan. 20, 1969.

to the final stages in the process of mechanization, but the effects of automatic methods of control have been ever-increasing in importance and, while some of these do not involve changes in labour requirements, others do. This is especially the case with extensive changes which affect whole processes or operational stages, giving scope for a major revision of labour needs. Such changes have also had their effect on the composition of the labour force, for automatic and semi-automatic or programmed plant require for their operation teams of specialist workers to set them up, feed them with information, service them, and make use of the additional data they provide. As a result, the labour-force structure for the industry as a whole has been becoming increasingly weighted by administrative, technical and clerical grades of worker, and in the newer plants this trend is dominant. Within the production grades, there is some suggestion that the 'medium-skilled' operatives are expanding proportionally at the expense of the unskilled man and of the worker with skills acquired by long experience, though this may well be more typical of the now largely completed phase of mechanization, than of the current trend towards more automatic control. But it would be wrong to attribute all the changes in the industry's labour-force size and structure to technological change, especially in recent years. Increasing attention is being paid by the industry to the need to reduce standard manning to levels more in line with those in other countries, and some companies had already made considerable progress in this direction just before renationalization.

CHAPTER 7

MANPOWER PROBLEMS OF ADJUSTMENT

I. INTRODUCTION

The previous chapter has reviewed the major trends of technological change in the steel industry, and has given some account of the labour-force effects of these changes. In this chapter our main interest is in the employment and industrial relations problems which have arisen because of technological change. We have so far said little about the problems of effecting change and of finding suitable methods for adjusting to new situations. We have said nothing about the part played by trade unions in regulating the pace or the form of change. We must consider the extent to which redundancy problems have arisen and how far companies have been able to soften its effects. What have been the effects on the skill requirements of the labour force? And what are the implications for labour in the next stage of development? These are the questions to which we now turn our attention.

The remainder of this chapter takes the following form. Section II examines the nature of employment problems which have arisen and the effect on adjustment of the seniority rule as it applies in this industry. Section III takes up the issues which commonly arise when technological change is experienced: the importance of demarcation, the development of manning standards, the evolution of wages and skill in the new technical situation, and the role of compensation and resettlement benefits. Finally in Section IV we take a brief look at some of the implications of nationalization.

II. EMPLOYMENT PROBLEMS

(i) *Some examples*
In the steel industry, the most significant feature of technological changes which are substantial enough to have important labour-force effects is that they usually take a good deal of time. The

164

process of planning, ordering, building and running in a major item of steelworks plant, such as an LD plant, a new rolling mill or a continuous-casting plant, is much longer than that of the ordering and installation of machines in lighter manufacturing industries like printing. An extreme case was the SPEAR project at the Steel, Peech and Tozer Works of United Steel, where new electric arc furnaces were built on a site occupied by fourteen open-hearth furnaces which had to be demolished to allow new building to commence. But this was not untypical of the steel industry, where most investment goes into existing plants which frequently lie in built-up industrial areas with space at a premium and extensions into new land not always easy. In the SPEAR case, admittedly an exceptional one in terms of its size and the problems of phasing involved to maintain adequate output from the outgoing open-hearth plant, the planning and construction period extended over a period of almost six years. In the general case, a time span of two to three years from the final investment decision until the commissioning of the plant (or the last piece of the installation) is by no means unusual, and this gives management two great advantages. First, they can draw up detailed plans well in advance, to cater for the phasing in of new productive capacity, the phasing out of capacity being withdrawn while still maintaining an adequate level of output, and the general engineering and production problems that inevitably arise. Secondly, and equally important, they can plan for the redeployment of workers whose jobs will be affected by the change. For redeployment, if it is to be successfully carried out with the minimum of disruption to production, morale and industrial relations, obviously requires an adequate time-span for remedial measures to be devised and gradually carried into effect.

In the SPEAR case mentioned above, 920 jobs eventually disappeared as a consequence of the changes. Even in a large works such as Steel Peech, this is a sizeable number, and a sharp reduction of this size would have presented serious problems. As it was, normal monthly wastage through voluntary leaving, retirement, etc., averaged about 30 workers per month, so that over a five-year period wastage could easily take care of the simple numerical problem. In practice, the matter was less easy. In order to maintain adequate production from existing plant, new workers had to be employed to replace some of those who left. Also, not all the workers previously employed in the two open-hearth melting shops would be required for duties on the new electric plant, and if they were not to be discharged, they had to be moved, as opportunities arose, to other

165

jobs which were both acceptable to them and approved by management. Again, the wastage figure of 30 per month included craftsmen, and since their employment was not threatened by the change, and the whole burden of redeployment was on process workers, the constraints on management were greater than might initially seem to be the case.

In the end, the company decided to make as much use as possible of normal wastage, yet declared a policy of no redundancy *for workers employed at the time the scheme was announced.* It was decided that, apart from craftsmen, new workers would be hired only on the understanding that they were employed in a temporary capacity. Thus, as established workers were moved on to the new electric melting shop, or to other jobs becoming available in the works, the essential gaps were filled by temporary labour. This policy met with a somewhat fortuitous success, for 1962–3 witnessed a steel recession, which meant that there were unemployed steel workers in the market, prepared to take such temporary work. This might have been much less practicable in late 1964 and 1965, when the industry was again working near capacity, but by this stage the need to recruit stop-gap labour had passed. The number of temporary workers varied throughout the scheme, up to a maximum of about 1,000: the average number at any time was around 500. In the end, about 250 were converted to permanent status, and 140 were laid off.

In this way, the impact of a major reduction in jobs was confined to a very small number of workers who had knowingly accepted temporary jobs, while all the established workers who wanted to stay on in the works were retained. The time available to management to bring about a gradual reduction in the established labour force was undoubtedly of considerable help, not only for planning, but also in allowing over a period a freer and wider choice to those workers who were not to be employed in the new melting shop. The final results of the complete redeployment exercise are shown in Table 7.1.

It is worth noting that although the company made extensive efforts to place men in other firms and especially in other branches of United Steel, this met with relatively little success. The Appleby–Frodingham branch was at this time short of labour and, although beyond reasonable daily travelling distance, was only about 40 miles away from Rotherham, so that a major move from the area was not involved. Despite organized visits to the Scunthorpe works, and the fact that housing could have been made available without serious delay, the response of the Steel Peech employees was small,

166

TABLE 7.1

Redeployment in the SPEAR exercise

Direction of Redeployment	Number of Men
1. Transferred to other branches of United Steel	34
2. Transferred to other companies	9
3. Transferred to other departments of Steel Peech	544
4. Retirements	151
5. Deaths	40
6. Voluntary leavers	140
Total	918

Source: Steel Peech and Tozer.

and for management, disappointing. The final results depended primarily on internal redeployment, raising further problems to which we return shortly.

Examination of several other cases of job redundancy, though mostly on a much smaller scale, suggests that this is a fairly typical approach. Companies have frequently relied on wastage over the time period of the change to minimize the impact, and internal redeployment has been commonly used. This assumes that the works itself is going to continue as a going concern. But as we have seen, the effect of technological change in steel has been to increase capacity, to reduce unit costs of production, and to lead to the withdrawal of older, higher cost capacity, which has led to a number of complete works closures. This is one of the indirect effects of technological change, resulting in a need for rationalization of particular sections of the industry or in geographical concentration. It is in such cases, where there is no possibility of internal redeployment, that the greatest difficulties of re-absorption have occurred, and it is worth while considering how such run downs have been effected.

Two cases can be briefly outlined. The first is the closure of the Blochairn Works, near Glasgow, by Colvilles in 1962. This works was mainly employed on jobbing, producing plate against small orders. Although the phasing out of Blochairn had been fore-shadowed many years before, it had suited Colvilles to keep it running, but in August 1962 a new four-high plate mill, with adequate capacity for all foreseeable needs and capable of producing a much higher quality product, was to be commissioned at Colvilles' Clyde Bridge Works. This was the end for Blochairn and its 320 works-grade and 70 staff personnel.

In preparation for this, Colvilles issued a directive to their other plants in the Glasgow area that new workers should not be recruited without first checking whether gaps might not be filled from Blochairn. For the 70 staff personnel affected, information on job history, age, service and current earnings, and addresses, was circulated among the other works. This measure was completely effective since by the end of 1962 all Blochairn staff were found new jobs within Colvilles, none of them experiencing a cut in salary. Of the works-grade people affected, 25 were of pensionable age and presented no problem. Twelve apprentices were transferred directly to other Colvilles' plants. For the remainder, details of age, experience, etc., were circulated to all Colvilles branches, and in the end all were made an offer of employment in one or other of these branches. In the event, many did not accept this offer. Blochairn was an old-established plant and had a high proportion of long-service employees who were strongly attached to the area. Most of the job-offers made to them were at Ravenscraig, which was then expanding, but although only 12 miles away it was not easily accessible by public transport from eastern Glasgow, especially for shift workers—and Ravenscraig was on continuous shift working. Another factor was that Colvilles (well before the Redundancy Payments Act of 1965) were to make resettlement allowances and severance payments, and some workers preferred to take these, leaving the possibility open that they might later find jobs in the Colvilles organization. In practice a small number did return to Colvilles.

In this case changes in productive capacity in one part of a large organization resulted in the closure of a subsidiary works. This involvement of a large organization provides the *opportunity* for men to be redeployed to other parts of the organization, but it raises more difficulties than in the case of redeployment among departments within a single works. In the SPEAR case, there seemed to be a preference for moves within the works rather than movement to other plants or other companies. In the Colvilles case what appears to have been a similar reluctance to take employment outside the immediate labour market area gave rise to more problems, for internal redeployment was ruled out—even though the labour market situation in the north and east of Glasgow at this time was poor, due to other factory closures and the prospect of a scheme of local redevelopment which would not immediately involve a replacement of factory premises and jobs.

The second case is the closure of the works of John Baker and Bessemer Ltd., in 1964. It would be misleading to attribute this closure simply to technological change within the steel industry:

it was principally brought about by a contraction in demand for railway products such as wheels, tyres and axles, which were produced by Baker-Bessemer, Steel Peech and Tozer, and Taylor Bros. Ltd. (a subsidiary of the English Steel Corporation). In February 1961, ISHRA approached English Steel and United Steel, suggesting they should buy Baker-Bessemer.[1] The two major companies stressed that they could make this purchase only with the intent of closing Baker-Bessemer, since there was vast over-capacity in that section of the industry. ISHRA, after consideration, proposed that such a run down might be phased over a two-year period, but this proposition was turned down since, at this time, the nearby Steel Peech works were embarking on the SPEAR project, which, as we know, was to result in the disappearance of 900 jobs. The problem of absorbing more than one thousand staff and wage-earners from Baker-Bessemer at such a time (when in any case there was a steel recession) was unthinkable, since neither company wished to be associated with a major addition to local unemployment. For the moment the issue was shelved, but in 1963 a review of capacity in the railway wheels and tyres section again stressed the existence of serious over-capacity, and the fact that even with the closure of Baker-Bessemer, further rationalization in the two major companies would be necessary.[2]

In the two and a half years since the purchase and closure of Baker-Bessemer had first been mooted, conditions had changed a good deal. The recession was over. The SPEAR project was well in hand and plans to avoid redundancy had been made. A new works at Park Gate was to be commissioned in 1964, and would be capable of absorbing up to 1,200 workers. And English Steel's Tinsley Park Works, opened in 1963, could take another 300 men.[3] In these circumstances it was fairly certain that the thousand workers at Baker-Bessemer could be absorbed much more readily than in 1961–2. English Steel and United Steel purchased the Baker-

[1] The Iron and Steel Holding and Realization Agency (ISHRA) was set up following the 1952 denationalization of steel to hold the assets of the Iron and Steel Corporation (the nationalized company) and to return them to private ownership as soon as possible on reasonable terms. Baker-Bessemer had never been returned to private hands.

[2] There was little doubt that the closure should affect Baker-Bessemer, and not one of the other two plants. Capacity in the latter was more modern; there had been a reorganization and replacement of plant in Steel Peech railway plant in 1962, and costs were lower.

[3] Park Gate was three miles away from Swinton, where most of the Baker-Bessemer employees lived, and Tinsley Park was twelve miles from Swinton.

169

Bessemer plant in October 1963, notice of closure was given to the Iron and Steel Board, and the closure announcement was made shortly afterwards.

The closure was thus timed to minimize the impact of redundancy and to phase in with expansion programmes elsewhere in the area. Management had three principal considerations in mind. Employees were to be given adequate notice: the closure was to be phased to suit employment opportunities and to prevent large numbers of workers coming on to the labour market simultaneously; and unemployment over the Christmas period was to be avoided. In the event, many left voluntarily before the end of the year or were found other jobs in the expanding companies. When the first phase of redundancy was effected, on January 4, 1964, 198 workers were made redundant, while a further 156 continued work until the second and final phase at the end of March 1964. One month after the date of the first phase, 159 were still unemployed. Four months after the second phase (i.e. the end of July) 80 remained unemployed. In the end 206 men of the total had been obliged to register as unemployed.

There are two points to be made here. First, one-fifth of the workers affected by the closure could not immediately find work, and more than one-third of these had a spell of unemployment that was more than transitional. The implication is that in a closure of this kind, where the firm is not an integral part of a large organization and where internal redeployment is not possible, the problem of re-absorption is more difficult. The second point is that the co-operation and activities of the companies in the area did reduce the problem to a manageable size: without these activities, the impact would undoubtedly have been more harsh. The management of Steel Peech and English Steel, in conjunction with Park Gate Works, carried out an intensive programme of interviews with Baker-Bessemer employees, with the result that about 400 were directly recruited by them, while smaller numbers went to other firms outside the steel industry by arrangements between the managements. The success achieved in this must be qualified by the number remaining unemployed four months after the closure. They appear to have been mainly in the 55 to 65 age-group, many of them with as much as 30 years service with Baker-Bessemer (and therefore entitled to large redundancy payments under the terms of the scheme administered jointly by the two principal companies). This might have been a case where retirement would have been a useful device, but it was not adopted. At least some of these unemployed were eligible for jobs at Tinsley Park and Scunthorpe, where vacancies remained, but the problem of immobility seems to have prevailed.

(ii) *The Seniority Rule*

On a more general level, the mobility of the production labour force in steel is seriously impeded by the seniority principle, and since this can have important consequences for the success of schemes involving redeployment within or between plants, it is worth while considering in greater detail. As observed in Chapter 5, promotion for process workers in the steel industry is almost entirely governed by seniority within a particular promotion ladder. Entry to any ladder is always at the bottom, so that when redundancy occurs, workers who find it necessary to move to another steel-works—even within the same company—have to enter at the bottom of the scale, and in consequence the drop in earnings may be severe. This institutional factor is one of the obvious causes of the resistance to geographical movement, quite separate from, and additional to, the normal psychological and economic costs of such transfer. This may help to explain why, in the last two cases examined, many workers were not prepared to accept vacancies offered to them even in plants within reasonable travelling distance.

However, the same considerations of seniority arise in cases of redeployment within a single plant. Mention has been made of the fact that many of the new technical processes are much faster in operation than the old. In steel-making as elsewhere this has meant that fewer units of plant are required. In such cases, the most skilled, highest paid workers on each of the old plant units cannot all find equivalent jobs on the new plant, and some of the first-hand melters (the top jobs on the ladder) must accept second-hand melters' jobs, with demotion carrying on down the scale and the short-service workers being ejected at the bottom. Elsewhere, and especially on the rolling side, there may as before be a single unit of plant, but now requiring fewer operatives. Here again there may be some demotion and transfer out at the bottom, but this is not always possible. Where continuous casting replaces primary rolling facilities, for example, the experience that is most appropriate is not that of the rolling mill crew but of the melting shop pit-side workers (from whom, for example, the Appleby–Frodingham continuous-casting crews were mainly selected). Other changes, such as the automatic inspection equipment introduced at Samuel Fox, meant that some groups of jobs disappear altogether. In consequence, there is always a likelihood that some workers, previously on relatively

171

high rates of earnings, will have to move to new jobs within the works at the bottom of the promotion scales. They cannot be pushed into the top or middle ranges of a rigid promotion system. A number of conclusions follow from these observations.

First, there may be considerable pressure on managements to select for the new plant men who are near the top of the seniority ladder on the outgoing plant, especially in cases where they cannot be pushed down a single grade but must, as it were, 'begin again'. Some managements have resisted this by insisting that some qualification must be obtained—for instance in electric steel-making where open-hearth practice gives way to electric—and other conditions are imposed according to circumstances. But the seniority rule, though not absolutely rigid, has been the over-riding factor in virtually all cases studied. Secondly, the downward reshuffling of workers in order of seniority on the ladder does have the effect of minimizing the degree of change for those who do have to move down. They will frequently be able to move to jobs which they have previously held on their way up the promotion ladder and which carry only slightly lower wages, while those who are edged out of the ladder at the bottom will neither have been earning the high pay associated with the more skilled jobs, nor be irretrievably committed to the acquisition of a particular type of skill and experience. For them, transfer to other work in the plant need not mean a serious loss of earnings nor the abandonment of an investment of time in acquiring a skill. Thirdly, the fact that *some* workers will have to move from high-paid to much lower-paid jobs as a result of the seniority system may have been responsible for the willingness of most managements to provide compensation payments as well as pure redundancy benefits.[1] Were this not done, the resistance both on the shop floor and from the unions might well be greater, and bargaining over the manning of new plant harder. Finally, the existence of a conventional rule accepted by labour, unions and management alike, may have eliminated some of the problems of adjustment. Where rules do not exist, or even where rules are drawn up *ad hoc*, there is much more likelihood of disputes between management and unions over the eligibility of certain workers for specified jobs.

On balance, therefore, we should perhaps not be too critical of the seniority rule, though *a priori* it would appear to be a factor militating against flexibility and the right of managerial selection.

[1] These payments were usually generous by the standards of the time, and redundancy payments were common before the advent of the Redundancy Payments Act.

Furthermore, when a new works is being opened up, the seniority principle cannot apply: thus in the recruitment of Baker-Bessemer workers to Park Gate and Tinsley Park, and also in the hiring by Spencer Works of surplus workers from Ebbw Vale and other neighbouring works, the problem of re-starting at the foot of the ladder was largely avoided.

III. PROBLEMS OF ADJUSTMENT

In this section, we look at four particular aspects of the problem of adjustment to technological change in the steel industry. First, there is the issue of demarcation which, though not a common one in steel, has on occasion given rise to serious dispute. Secondly, there are the procedures adopted for the determination of manning on newly introduced plant. Thirdly, in many ways closely related to the manning issue, there is the determination of wage levels and structure in a new technological environment in which skills and job content may have undergone substantial change. Finally, there is the approach of the companies to compensation and resettlement.

(i) *Demarcation*

Two demarcation disputes are particularly relevant. The first has made several appearances since 1958, when the Steel Company of Wales (scow) proposed to use Confederation members to fix dolomite block linings into new vln converters. This job was claimed by the aubtw (the bricklayers' union). Discussion produced no decision and the company adopted a different practice with no dispute. In 1962, when ld plants were growing in importance, a similar dispute arose between the same unions and an arbitrator was appointed. Both parties were keen to obtain this work since both were suffering employment declines due to changes in technology[1] and the new work on ld converters was a means of stabilization. The award by Professor (now Sir Daniel) Jack favoured the Confederation. After a successful intervention by the tgwu at Consett, on the basis of a local agreement, the aubtw reopened their case, but two claims by this union were rejected by the Industrial Court.

Meanwhile the scow management had made several efforts to reopen the discussion of 1958, without success. But in 1966 the Company announced its intention to adopt large scale ld practice,

[1] Mention has been made of the decline in production employment. For the bricklayers, the problem was the reduction in open-hearth furnaces, the relining of which provided them with regular employment.

and its proposal to use dolomite blocks which were to be fixed by Confederation members. The AUBTW threatened strike action and the matter was referred to a Committee of Investigation appointed by the Minister of Labour under the Conciliation Act of 1896. The finding again favoured the Confederation.

This was principally a dispute between the AUBTW and the Confederation. The basic arguments of the AUBTW throughout were very similar. Fixing dolomite blocks was, they alleged, a skilled job, requiring bricklayers' tools. The safety factor was also introduced by the AUBTW on the ground that structurally unsound linings in converters could cause damage and personal danger. The Confederation remained rather aloof after the Jack Award, taking the view that the AUBTW was bound 'legally and morally' by that Award. It had not itself ever 'sought to oppose the introduction of new techniques, because it had recognized that to do so would be to oppose the attempt to make British industry more efficient and to maintain and improve the standard of living'.[1] The Committee of Investigation, set up in 1966 under the chairmanship of Professor D. J. Robertson, tried to clarify the position once and for all. It reaffirmed the correctness of the Jack Award, observing that the task in question was repetitive, requiring no high degree of accuracy, and forming a continuous operation from the stripping of the old lining to the installation of the new.

The fear of the AUBTW about declining work for bricklayers in the industry has already been remarked upon. But the Confederation was in a similar position for, as we have seen, the effects of technological change on production operations and employment has been marked, with a sharp decline in labour requirements in many sectors. Under these conditions the Confederation's behaviour was exemplary, in that it had not resisted change. The Confederation, we might conclude, had simply tried to secure the best possible terms for its members whose jobs were affected; and indeed, since many of those who were so affected were relatively unskilled, it was reasonable for the Confederation to try to obtain or retain a hold over work which was suitable for them.

The second case of note is related to the running in and building up a new plant producing stainless steel 'hollows' (long hollow cylinders) in the Stewarts and Lloyds Tube at Corby. The company entered into an agreement with the Confederation concerning the

---

[1] *Report of an Inquiry into a Dispute between the Steel Company of Wales and the Amalgamated Union of Building Trade Workers*, Ministry of Labour (HMSO), 1966, para. 39.

manning and wages of a machining operation on the 'hollows'. This operation was part of a newly developed method of producing stainless tubes, and the issue of the allocation of this work had therefore not arisen before. By October 1965, six of the machines had been installed and were manned by Confederation members, but, due to the need for further development work, three turners from the Research Department, all members of the Amalgamated Engineering Union (AEU),[1] were brought in. In January 1966 they were replaced by Confederation members, but in March the AEU claimed that the machines were lathes and should therefore be manned by AEU members. The Company argued that this was an essential stage in production and that production work was conventionally manned by Confederation members.

A number of complicated issues now developed. The company and the AEU finally agreed to the principle of sharing the work between the disputing unions but the Confederation refused to accept this solution. The AEU withdrew their labour when two further machines were introduced on an operating basis, manned by Confederation members, and an official strike began on January 31, 1967. Attempts to resolve the dispute by the TUC were unsuccessful and a Court of Inquiry was appointed to investigate the case, finding in favour of the work continuing to be done by the Confederation members.

The interest of this dispute lies in the fact that the new process involved the use of a machine in the course of production which was perhaps—though the parties would not agree on this—more closely related to a turning operation in engineering than to a normal steelworks production operation. To quote the Report of the Court of Inquiry:

'In the view of the AEU the operation of centre-lathes was the sole prerogative of members of the union trained as turners, who might be skilled or semi-skilled according to the nature of the work undertaken. The right of members of the AEU to carry out this work was not affected by the degree of precision called for, or by the purpose for which the work was undertaken. Because of its fundamental importance, the union could not agree to allow the right of its members to operate centre-lathes to be the subject of any form of arbitration or, as the Company had sought to do, to be ignored.'[2]

[1] As it then was.
[2] *Report of a Court of Inquiry into . . . a dispute . . . at Corby* (Cmnd. 3260, HMSO), May 1967, para. 18.

As in the dolomite block cases, then, the union—in this case the AEU—was concerned with the rights of its members to carry out operations requiring the use of a particular kind of tool or machine; whereas the company and the Confederation adopted the view that the duties on this machine constituted an integral part of the production of a steel product, and as such belonged rightfully to the Confederation's workers. Admittedly, although as a general rule in the industry 'production' work tends to be done by Confederation members, there are important exceptions, and according to the Iron and Steel Trades Employers' Association (ISTEA) representing the company in the Corby tube dispute:

'no union in the iron and steel industry had established the sole right for its members to perform any particular type of work . . . [and] the rights of members of unions to undertake particular kinds of work had to be based on either specific agreement or on custom and practice.'[1]

In any case, even if there were such a rule, agreed upon by all parties, there may continue to be scope for disagreement over what is production work and what is not. It is here that the wider implications of this case become apparent. The continuous advance of technology in the steel industry, and particularly the extension of the range of products being produced by the industry involving the use of new methods of finishing, is likely to increase the degree of overlap between what have traditionally been accepted as processing operations, and tasks which are more akin to the work normally done on finished steel products by workers in mechanical and electrical engineering works or departments. The relative infrequency of demarcation disputes in the past has been due largely, it seems, to the existence of conventional and customary local or works agreements about the meaning of production and non-production work, coupled with the presence of the Confederation within which there is no real scope for internal demarcation. So long as the type of processing operation in the industry remained within certain traditional limits, and the product range was clearly within the normal scope of the steel industry *per se*, the incidence of demarcation issues were likely to be small. But in the present circumstances with the traditional boundaries of the industry expanding and changing, the potential for conflict on this score may well be on the increase, particularly between the Confederation on the one hand and the AEF and ETU on the other.

1 Ibid., para. 24.

Furthermore, the recent and continuing contraction of manual employment in the industry as a whole and in some sectors in particular is the kind of situation in which all unions will become concerned about their own members' employment, and seek out areas where production is expanding as a means of stabilizing their own membership. On this count, too, the industry may have to anticipate demarcation problems as technological advance continues and employment declines further.[1]

(ii) *Manning*

The discussion of manning here relates not to the allocation of work to members of a particular trade or union, but to the development of a manning practice for a plant or process, which can be regarded as a 'standard'. In the steel industry there are relatively few major units of plant which are identical from works to works, most of them being 'one-off' or at least varied substantially from a basic model. This, and the dominance of workplace bargaining, has resulted in individual manning standards which are both the product of the way in which management has approached the manning question, and the outcome of subsequent negotiations. Even for very similar plant, as measured by appearance, rated capacity and actual performance, standard manning may differ from works to works, often because of the presence in some plants of ancillary equipment but sometimes because of differences in the total organization of work within a department: both these elements will be reflected in negotiated manning.

The evolution of a manning standard will typically take the following form. Management generally assumes initial responsibility for drawing up manning lists, which are then presented to the unions concerned as a basis for negotiation. Management's manning lists are usually based on assessments by work-study teams, necessarily prior to the completion of the plant itself. Nevertheless, information from the plant manufacturers and the company's knowledge of its own specifications for the plant will allow work-study experts to

---

[1] Other demarcation disputes have occurred but these two seemed most relevant to the issues involved here. It should also be said that in these two cases there were many more questions involved than is apparent from this brief account. For further details, see Industrial Court *Reports* and *Awards*, Nos. 3026, 3054, 3055, and the *Report of the Inquiry into a dispute between SCOW and the AUBTW*: and the *Report of a Court of Inquiry . . . at Corby*, Cmnd. 3260, quoted above.

schedule a normal routine of operation for the plant, and the frequency of particular duties. From this can be built up a provisional picture of the manning required, and more intensive work can be begun on job description, the evaluation of jobs and subsequently management's views on the level and structure of wage rates. More sophisticated methods are now being used. For example, the manning for the new addition to the electric melting shop at the Samuel Fox Works was worked out by computer from a simulation programme, enabling management to assess the frequency of tasks over a (simulated) two-year period, allowing for a variety of normal and abnormal conditions. In addition, of course, company representatives will often visit other works, in this country and abroad, which are using similar plant.

These management estimates are subject to preliminary negotiation with union representatives, pending the development of management proposals for wage rates and methods of selection for the new plant. On the evidence available from our case studies, agreement on manning is usually reached within the works, i.e. at the first stage of the normal negotiating procedure. The single exception found was in the SPEAR case, where the Confederation pressed for four melters per shift to be used on each furnace, rather than management's proposal for three. Union representatives claimed that four men were required to achieve the scheduled production cycle of three hours, as opposed to eleven hours on the old open-hearth furnaces. Management's work-study estimates, backed up by comparisons abroad, indicated this schedule could be achieved reasonably by three melters. In the end, this dispute went to the National Joint Sub-Committee[1] where it was settled in favour of the company.

Of course, such advance provisions for manning and job descriptions must be subject to some adjustment. Certainly, the estimates can become increasingly educated, but the more detailed the resultant job-descriptions proposals become, the more open they are to differing interpretations and the more common do wage claims, based on minor changes in practice, become. In addition, the running-in of a new plant requires a good deal of on-the-spot development work and there is no doubt that in the past some of the additional workers brought in to cover teething troubles have 'stuck', and the standard manning originally negotiated becomes expanded by custom and practice. It is arguable that at least part of the over-manning now existing in some parts of the steel industry

[1] See Chapter 5.

is due to just this expansion. Hence it is increasingly recognized that realistic manning standards must, if possible, be set and adhered to from the outset.

Two factors reinforce this need. One is that the negotiated manning standards and wage rates form part of a works agreement, which can normally be departed from only on a change of practice They are, therefore, relatively (though not absolutely) rigid, especially for downward adjustments. Secondly, there is the existence of the seniority system, which makes the reduction of manning more troublesome, since those who are removed from a process may have limited opportunities elsewhere in the plant at a comparable rate of payment. For many years, the scope for management's activities to reduce standard manning was regarded as relatively restricted, but more recently there have been a large number of works-based agreements of the 'productivity' type,[1] which have reopened the question of manning-payment relationships which once seemed unalterable in the context of a given technology. Other developments have included the national negotiation of a 40-hour week for the industry on the condition that the overall cost of labour to the companies will not rise. This implied at least a 5 per cent cut in labour, and works could not move to a 40-hour week *until* this reduction had been achieved. This unusual agreement, though not related to technological change, was nevertheless a sign that the industry was making a serious effort to improve its utilization of manpower, and that in itself may be symptomatic of the attitudes to manning likely to be adopted in the technological conditions of the next few years. Moreover, it may be that the current phase of productivity bargaining will do as much or more to change the level and structure of employment in the industry than technological change itself.[2]

Manning, therefore, is an important area of collective bargaining in the steel industry, as elsewhere. But as in other industries, the question of manning is related to the wages for particular jobs, and the two issues of manning and wage payment are often part and parcel of the same set of negotiations.

[1] These are often very small in their effect, reducing manning standards by two per shift here and five per shift there: but the cumulative effect of these, over a period of one or two years, is quite marked. This process is continuing under nationalization. In the *Annual Report* of the British Steel Corporation for 1967–8, it is recorded that agreement on at least 15 individual plant productivity agreements was reached in a six-month period.

[2] Cf. the discussion in Chapter 13 below.

179

(iii) *Wages and Skill*

Once management has evolved its own ideas on manning, it will begin to develop job descriptions providing the basis for the wage structure on the new plant. The precise form of this exercise will vary from case to case, but in some cases job evaluation techniques area dopted,[1] including allowances for skill (as measured by degree of training and experience required), responsibility (including allowance for the value of tools and materials employed, and the extent of control over the process), and smaller allowances for physical effort, working conditions and possible hazards. By this means it is possible to determine the structure of what will eventually be the wage system, but before the wage system can be properly worked out, two further matters have to be taken into account—the absolute *level* of payment, and the *method* of payment. Both these points are of particular importance in the adjustment to technological change.

By the absolute level of payments here we mean the monetary equivalent of the work units measured from the job evaluation. If one job is valued at 100 units of skill, responsibility, etc., and another at only 80 units, this does not tell us how much a worker doing either job will be paid. Most firms concentrate initially on the payment to be made for the job which carries the highest evaluation within the process,[2] though of course even if management have an initial view on what this payment ought to be, the actual outcome will depend on negotiation. Once this is decided, the rest of the wage structure usually follows from it directly. The determination of the top rate, of course, cannot be determined in abstraction. Not only does it have to take account of the earnings levels for similar types of labour in the rest of the industry and in the external local labour market, but it is usual for managements to consider how it should relate to the rates previously paid on plant which is being replaced or on jobs from which the new labour force is being drawn. It is seldom possible to pay less on the new plant than was being paid on the outgoing plant, and since new plant will

1 Not all steel companies used job evaluation, but in one which did it was seen as an integral part of establishing a rate, first because it provided a firmer basis for negotiating comparative rates, and secondly because it was felt that the nearer the rate is to representing the actual work of a job, the greater the chance of having a satisfied labour force.

2 Where a whole new plant or works is involved, there will also be the problem of determining the relationship between the top-paid jobs in each of the various processes or sections.

almost invariably have a higher rated output and frequently a lower labour complement (implying higher labour productivity), management generally set a rate of pay rather higher than on the old equipment. In the steel industry, the top rates of earnings for any job are set in line with the expected normal output rate of the plant, by means of a production bonus, but as we see below problems may arise due to difficulties in running-in the new plant.

The second point of note is the method of payment, which invariably raises the question of the bonus element in pay. As mentioned in Chapter 5, tonnage bonus payments constitute a sizeable element in production workers' earnings and in some cases, as under the old Melting Rates Agreement, it was almost the sole constituent. In the past this has been justified by the contention that in certain processes a first-class team could directly influence output from their plant to a very significant degree, and the tonnage bonus element thus operated as a strong incentive to higher output. As a result, the tonnage bonus is a traditional feature of the steel industry wage system, from which it is now proving hard to break away. Many managers now argue that recent technological developments have made it much less practicable for workers to have any sizeable influence over output, since so much is done automatically, and since so much instrumentation now exists that the element of discretion and judgement has diminished in importance. Though it is by no means true that the plant 'runs itself' there is less scope for individual workers to improve productivity above the set target, and some managers argue for a straight weekly rate of pay, though others would go less far and settle for a smaller tonnage payment element of around 20 per cent of gross weekly earnings. This, they point out, would mean greater stability in steelworkers' earnings but this approach to incentives is not one which has been received sympathetically by union negotiators.

It is difficult to reach a straightforward conclusion on this issue. In the first place, the real situation is often obscured by the necessary resort to an interim rate of payment while the plant is being run in. During the first six months or so after commissioning, it is often necessary for further development work to be carried out. The elimination of teething troubles means that the plant is out of action for parts of a shift or more. Engineers and technicians from the firm responsible for constructing and installing the plant are often present, and other temporary assistance may be brought in from other departments. The plant crews themselves will still be learning how the equipment behaves under production conditions. These factors, and the need to give workers a reasonable wage during the run-in period, may result in the setting of interim rates which normally comprise a guaranteed wage rather below the final scheduled level of earnings.

While it would be invidious to mention actual cases, the actual running-in time frequently exceeds management expectations. Exactly where the fault lies is not clear. Sometimes the plant itself has been at fault and technical difficulties have extended well beyond the expected run-in phase. In at least one instance the view has been taken that the failure to reach rated levels of output arose from the lack of incentive built into the agreement covering 'normal' production rates—though it is hard to decide here whether the interim rates were too high or the 'permanent' rates too low or insufficiently progressive. This view would imply that workers, even on highly automatic plant, can still influence output. But the variety of experience of different firms suggests that the degree of influence is unknown in advance and there is an ever-present risk that the initially agreed rates will have to be modified on the basis of experience. Still more difficult issues arise when the plant over-fulfils its rated output or fails over a period of two or three years to live up to its specified performance: examples of both tendencies have been encountered. But in the case of overfulfilment at least there is inevitably some doubt about its cause: whether it is due to the plant or to the efficiency of the crews who operate it. Other factors have complicated the interpretation as for example when a recession in demand causes output requirements to fall during the run-in period. Management does not want full capacity working and labour has no incentive to reach maximum output levels so that attitudes and work-practices may become set in this mould and the resulting inertia of custom and practice may prove difficult to overcome.

Another factor which bears on the relationship between rates on outgoing and incoming plant is the change in skill and job-content. The concept of skill is complex, and is perhaps more properly a subject for analysis by the psychologist rather than the economist, though it has important implications for the latter. A number of skill components can however be identified in the steel industry, and something can be said about the way technological change has affected them.

(a) *'Technological' skill* may be regarded as the knowledge and experience connected with the particular technology of production. One of the best illustrations of this is the different technological skills required in the melting-shop teams. Open-hearth technology demands considerable perceptual awareness on the part of the key workers, who have to know the various stages through which the charge will pass, to be on the look-out for danger signals, and make critical decisions about tapping-time, additions to the charge, etc., on the basis of direct observation of furnace conditions. But this steel-making process is long and slow, compared with newer techniques of oxygen and large-scale electric steel-making, which are based on quite different technological principles. The perceptual skills of open-hearth steel-making have largely given way to simpler observational duties as a result of improved instrumentation. Yet at the same time, because of the greater speed of the process, more anticipation of events is required from the crews.

(b) *'Operating'* skills are those which are related directly to the physical lay-out and operation of the plant itself. In general, the new steel-making and steel-processing techniques are much more complex than the old, with greater instrumentation, more controls and more information to be assessed, absorbed and acted upon.

(c) *'Communication'* skills relate to the need for teamwork and liaison both within and between plant crews. Under the older types of technology, the teams tended to be close-knit and in close contact with one another, where commands could be given directly, and each man could see for himself the conditions which required certain action. In steel-making, continuous casting and rolling mills, the members of the team depend at least as heavily as before on communication, but two changes have occurred. First, the teams are less clearly in physical contact, for as the size of plant has extended, as automatic or other controls have been introduced and men replaced by mechanical or other aids to production, the teams have become more spread out. Secondly, in many instances the teams have much less direct contact with the process itself, since instrumentation has taken over much of this function and the men can be placed in more comfortable working positions. As a result, operatives increasingly have to take decisions on the basis of second-hand knowledge, communicated to them either by instrumentation or by members of their teams whom they cannot see. This situation calls for greater precision in communication, confidence in instructions and information received from instruments or inter-communication equipment and, in the end, greater psychological stress. Many companies have introduced closed-circuit television and mimic diagram systems to bring a visual element into the work.

(d) *'Physical'* skills are, strictly, not skills at all, but rather

183

characteristics required in workers who have to manipulate heavy articles and work in very hot and noisy conditions. Technological change, in steel as elsewhere, has brought about major reductions in these requirements, though they are by no means entirely eliminated: heat and noise especially remain. Skill change is therefore not a readily quantifiable concept, and even if the various constituents of skill change were measurable there would still remain a major problem in knowing what weight to attach to each in converting them into money-wage terms. Perhaps the most that can be said here is that technological change in steel has brought about considerable changes not only in the skill structure of the labour force as a whole, by raising the ratio of administrative and technical personnel to production workers, but also in the work-content of production workers themselves. The latter are increasingly required to carry out their functions more rapidly and with more anticipation, and have to do so on the basis of information which is different in kind and degree from that to which they were previously accustomed. Whether in the last analysis this skill is greater or less than the old is indeterminate and perhaps is not a critical question. What does matter is that workers who have to move between jobs involving different characteristics—operational, technological or physical— must be capable of making the adjustment. This makes the selection process for the key jobs on new processes a vital activity, and since the best men available have to be picked, if possible, it is inevitable that the earnings level for this work should be *at least* as high as that on the old work, to provide the necessary attraction.

Another aspect of this kind of change in skill requirements is that the industry as a whole now has to pay more attention to the kinds of labour it is recruiting, and the wage that is relevant to recruitment needs. The work involved in the new techniques being employed in the steel industry requires qualities which were not necessarily present in the traditional labouring grades from which promotion lines were continually replenished. Thus so long as the traditional system of promotion continues, it is essential that the pool from which such key workers are drawn should be composed of men with the appropriate qualities. A number of companies, recognizing this problem, have put increasing weight on recruiting young workers and introducing them to the industry by means of what is virtually a three-year apprenticeship scheme, at the end of which they have a range of experience and an adequate background of technical knowledge. At the same time, the jobs in the labouring pool, into which these trainees are then fed, have been restructured

184

with the dual aim of making them more varied and interesting, and better paid. In another case, the manning of the new Spencer Works, a high proportion of labour with no steel industry experience was recruited and, although in the end this labour did not fill key jobs, the possibility of such a policy, with short intensive training courses, remained open until a late stage in the planning. Thus the needs of the changing technology are becoming increasingly reflected in the training and payment system of the industry.

A related issue is that of the wage-relationship between craft and production workers. The top-paid production workers have always enjoyed a substantial differential over the top-paid craft worker employed on maintenance, and this has long been a source of friction in the industry. Neither the existence nor the size of the differential have been noticeably affected by technological change, but two points of interest have arisen. One is that with the increasing sophistication and complexity of modern equipment, and the serious effects on output which can result from a breakdown, some managers would like to have available, at craftsman and technician level, a more specialized, higher grade craftsman or technician who would be used for specialized diagnostic purposes rather than routine tasks, and who would be paid according to this responsibility and skill. Yet most of the craft unions are geared to a payment system whereby all journeymen receive the same level of payment, aside from small allowances for seniority. Secondly, there is some evidence of an argument on the craft union side that, although a differential between top-paid process and craft workers is justified, the number of process workers receiving higher wages than craft workers should be much more limited. In the negotiations connected with one case of technological change, the craft unions proposed that no more than 4 per cent of process workers should be entitled to receive more than craft workers (which would presumably mean the senior men on each process) and that all time-served craftsmen should receive a wage no less than any of the remainder. It seems certain that more will be heard on this issue as technological change proceeds, for in some sectors of the industry at least the responsibility of the process worker is increasing less quickly than that of the maintenance craftsman and, as in the demarcation discussion, further changes can be expected to bring the two groups into increasing contact and perhaps conflict.

The role of wages in the adjustment to technological change, then, is complicated and involves a great many problems of different types. While this is by no means a full discussion of wage issues in the steel industry, it does provide some idea of the relevant points which arise in practice.

(iv) *Compensation and Resettlement*

As in manning and wages policy, management normally reserves the right to make the first proposals on the selection of men to operate the new plant, and to decide what policy should be followed with respect to those who either lose their jobs or suffer a reduction in earnings. The selection process has to be begun at a fairly early stage of the overall implementation of the change, and frequently seems to be started upon as soon as the general outline of jobs on the new plant are formulated. This is certainly true of the key jobs, which are likely to give rise to the greatest problems. The men who are to occupy these jobs will often have to undergo a period of training, sometimes including spells abroad or with other firms. Again, those who had key jobs on the old plant, but who cannot be fitted into similar jobs on the new plant, will have to be found new jobs—and plans made for their transfer and, where appropriate, compensation for redundancy or reduced earnings.

The process of selection for the new plant, though not rigidly determined by the seniority system, is closely governed by it, and in this respect the firm's task is made somewhat easier. Among the firms interviewed in the course of the study, the general procedure has been for management to make out a list of the workers for the new plant, and present it to the union representatives concerned for checking. In this way any obvious anomalies in selection (due for example to doubts about an individual's place in the order of seniority) are eliminated before the lists are made public. Subsequently, there may be an Appeals Committee, including management and union representatives, to deal with complaints by individual workers.

At this stage it is possible for management to begin to reserve jobs elsewhere on the plant for displaced workers, and a policy for compensation can be evolved. In practice both these areas, so far as most of the steel industry is concerned, were found to be very much the preserve of management. Advice might be sought from relevant union representatives and information given, but it was not an area in which joint negotiation was carried out. This is, after all, a phase of the exercise to which, because of its delicacy, managements will pay great care and attention to detail, and their policy must be seen to be fair over the range of individual changes— though it can hardly be expected that everyone will be happy. There is also the point made that before the introduction of the Redundancy Payments Act in 1965 there was no compulsion on firms to pay redundancy compensation, while after its passing into effect there was no more than a mimimun standard which firms had to meet. Nor is there any real compulsion on the firm to pay compensation for reduced earnings in jobs elsewhere on the plant. Despite this, it appears to have been fairly widespread practice in the steel industry, well before the implementation of the Redundancy Payments Act, to give compensation to men rendered redundant by technological change. And even although there is no legal obligation on firms to make compensation to workers remaining in employment on reduced levels of earnings, it again seems not uncommon for such payments to be made.

The questions which then arise are how the firms viewed these payments, how they worked out appropriate levels of compensation or allowance, and how the unions and workers reacted. In the first place, there is no doubt that the firms regarded it as a moral obligation upon themselves to make the transition for the displaced workers as comfortable as possible, though in the earliest cases the greatest attention was paid to the longer service workers and the older workers (who are often one and the same). There are always other motives which might be imputed, of course, such as 'sweetening' the union's attitude to change, or reducing the bargaining conflict over wages and manning on the new plant, but so far as can be determined, the adoption of a compensation policy was primarily due to unselfish motives.

Four forms of compensation were found: the straight *redundancy payment*, to be paid on severance of the worker from the company; the *supplementary unemployment benefit*, an allowance paid on severance over and above the national insurance benefit to those unable to find alternative work; the *transfer grant*, paid to workers who move to other firms or branches by arrangement with the original employer and who are thereby involved in removal and settling-in expenses; and *reduced earnings compensation*. The basis for each of these, it will be noted, is rather different. The first is a reward for service: the second a payment to enable displaced workers to exercise a wider choice, spread over a longer period, among jobs available in the local market: the third a sum to cover direct costs incurred in transfer; and the fourth a payment to cushion the effects of a marked drop in earnings. Because of these differences, it is not unusual to find firms paying out more than one kind of allowance. In the closure of Colvilles Blochairn Works and the

187

Baker-Bessemer Works, redundancy payments and supplementary unemployment benefits were made. In the SPEAR exercise, there was only a reduced earnings compensation payment, but this applied not just to those who stayed in the works at a lower rate of pay: workers who refused the offered job and left for other employment were paid compensation on the basis of the difference in pay between their old job and the job they were offered at Steel Peech and Tozer. This principle also applied to those who transferred to other United Steel branches.

No attempt is made here to cover the details of the various schemes employed. The main factors taken into account have, however, been the now conventional ones of age, length of service and normal weekly earnings. A formula to include these was worked out, and applied in the individual cases, usually subject to some maximum set by the company. At this point both the Employers' Association and the unions have had occasional influence, since it is in the interest of both to try to exert a co-ordinating function to keep firms roughly in step with each other. The Redundancy Payments Act has made for much more uniformity in severance payments but there is still no obligation to pay other forms of compensation, and the Confederation at least would seem to welcome the prospect of further legislation or a national industry agreement on this aspect.

Finally, there is the question of worker reaction. Mention has already been made of the general reluctance of men whose jobs are becoming redundant to transfer geographically, even to nearby branches of the same company. What is perhaps more surprising is the tendency for some workers facing the prospect of redundancy to refuse to believe that they will really be permanently displaced. In the Samuel Fox case, when the new stainless mill was introduced, there was a widespread belief on the part of the operators on the old mill that the new mill would not be able to produce a product of sufficiently high quality, and that their jobs would therefore remain. In other cases, workers have accepted inferior jobs in the redeployment process, rather than move to other firms where better paid work was to be had, in the belief that the company had underestimated labour requirements on the new plant and that they would be recalled to higher positions. Only very occasionally have such beliefs come true for a few workers.

Two other points are worthy of note here. One is that operatives on a plant that is to be withdrawn, for whom only lower grade jobs will be available after the change, have tended to prefer to remain on the existing job, with high earnings, for as long as possible. This is so even though it may in the end mean accepting a worse job than could have been obtained by moving earlier, when the choice was wider. Once a move has to be made, however, many of the workers, especially in the older-age groups, have not been unwilling to accept lower-paid jobs, which usually mean less arduous work and less responsibility, for which they are then prepared. The younger operatives, on the other hand, have been more ready to move earlier and so begin to move up a new promotion ladder as soon as possible. In other cases they have been more willing to leave the company or to transfer to another branch, where a new start can be made, often in a situation where promotion will come more quickly due to labour shortages. The second point is that, as in the Baker-Bessemer closure, a small proportion of the affected workers have seemingly made little effort to get new employment, even though opportunities appear to be available. Some may have regarded the redundancy as an opportunity for early retirement, especially where they had long periods of service behind them and substantial compensation. But there do seem to be others who hold out hopes of being able to find jobs comparable with those previously held, or who have some expectation of new employment coming into the area—this is certainly true in the Baker-Bessemer case[1] and to a lesser extent in the Colvilles closure of Blochairn.

Although such problems have arisen in the course of technological change and rationalization in the steel industry, they have been small in relation to the number of jobs affected. The main priority of the companies has been avoidance of redundancy of established workers, and a great deal of emphasis has been placed on the need for internal redeployment, coupled with carefully worked out and well-administered compensation schemes of several types. In works which are employing several thousand workers, the scope for such redeployment, and the avoidance of worker redundancy by policies of only minimum replacement of wastage, is much greater than in the small firm. It is also important that many large companies, with substantial resources in management and in financial terms, have

[1] This seems to have been the only case where a major dispute has arisen over redundancy, and this was a curiosity arising out of the regulations made for compensation on the denationalization of steel in 1953. A long legal battle ensued, the Court of Appeal finally ruling that had Baker-Bessemer not been transferred from ISHRA to new ownership it would still have been a going concern. Though the companies had originally paid fairly generous compensation to those made redundant at Baker-Bessemer, they were obliged to pay more. But this case was very much an exception.

189

been involved. Yet if, as now seems to be the case, steel companies are increasingly seeking and finding ways of reducing standard manning below previously accepted levels, by productivity bargaining and similar devices, employment in steel will begin to shrink as a result of the two separate influences of technological change and the standard manning reductions. In this respect the slower growth of product demand will not provide much of a counter-weight. This *need* not mean an increase in redundancy of workers already in the industry, but it will mean a contraction of jobs available in the industry. And since the industry tends to be heavily concentrated in certain areas, and is frequently located in towns which have few other sources of industrial employment,[1] the changes in labour demand may have considerable repercussions. A prolonged continuation of the kind of changes just mentioned—and these seem probable rather than possible—would seem to imply a need for some co-ordination of company policies with government plans for area development, industrial transfer and labour mobility. From this point of view at least, it is perhaps appropriate that at this time the industry is starting on a new career as a nationalized concern, where the scope for such co-ordination may be much greater than before. This, then, is the final question to be taken up.

IV. NATIONALIZATION AND RATIONALIZATION[2]

The Royal Assent to the 1967 Iron and Steel Act was given on March 22, 1967, with vesting date being fixed for July 28. As a result, the fourteen principal steel companies and their many subsidiaries became nationalized, leaving behind a large rump of smaller, private firms mainly concerned with re-rolling activities. The nationalized companies, in 1966, had about 268,000 workers and a crude steel output of 22 million tons. Almost immediately after vesting date the first Report[3] on the organization of the British Steel Corporation was published, revealing that there were to be four main Groups, based mainly on broad geographical lines but with attention being paid also to product grouping, the basic unit being the works rather than the previous operating company.

[1] E.g. Port Talbot (scow) and Ebbw Vale (rtb).
[2] This discussion relates only to a few aspects of nationalization which have a specific bearing on the question of technological change and employment. The views expressed are not to be attributed in any way to company personnel who assisted in the provision of case-study materials.
[3] Cmnd. 3362, HMSO (1967).

Nationalization is not likely to change the view expressed on a number of occasions by the Iron and Steel Board towards the end of its life that no new greenfield integrated works will be necessary for several years.[1] The problem for the immediate future is not expansion of capacity but rationalization to make the best use of existing plant. But rationalization can mean a variety of things. The new organization already outlined, together with the affirmed short-term aim of reducing costs, and especially fixed costs, would seem to imply a pooling of orders within Groups. The previous dispersion of orders from customers to individual works meant that production runs were shorter than need be, especially in rolling. In consequence, frequent changes in roll sizes, metallurgical composition, etc., had to be made, leading to plant standing idle at frequent intervals and output lying well below possible capacity. Of course the present world surplus of capacity makes it unnecessary for maximum outputs to be achieved, but it also means that markets are extremely competitive, so that steel has to be produced as cheaply as possible. If orders are pooled and allotted in such a way that longer runs can be achieved, unit costs will be cut, and it is this which is at present the main incentive for pooling.

This can, however, be only a first step, leading to others of a more radical nature. Mention has already been made of the considerable economies of scale which exist in the steel industry, and the fact that they seem to be obtainable at production and capacity levels far above that of any British works at present. It may well be, then, that for each of the Groups, certain operations will be expanded in scale so that they can cover the needs of the whole Group. An obvious example might be the use of a single melting shop (or certainly a more limited number) to provide the ingot steel requirements of a whole Group, using, for instance, much larger electric furnaces or LD converters and so deriving the available scale economies. This kind of development will mean that existing sections of plant will have to run down and close, and in the longer run it could mean complete closures of works. In either case, employment will be affected adversely. And this in fact seems to be the key to the future shape of the industry. 'Rationalization' in many cases can only imply a longer-term policy of closure and the concentration of production into larger units through intensive rather than extensive capital investment to derive the available economies of scale. There is little doubt that if the British industry is to remain a competitive

[1] This view has now been reaffirmed by the British Steel Corporation in its Report for 1967–8, p. 17.

191

force in world markets, and if some form of protection for the home industry is to be avoided, rationalization of this kind must take place.[1]

As preliminary statements from the British Steel Corporation made clear, a nationalized industry pursuing a policy of rationalization by closing down some plant and concentrating production into selected areas is faced with severe political and social problems. Political or social criteria, not economic calculations, were at the root of the decision to build two relatively small strip mills— certainly much smaller than would be warranted by purely economic considerations—at Newport (RTB) and Ravenscraig (Colvilles) in 1960–2. Both these sites were in Development Areas where unemployment was relatively high, and pressure for new jobs in these areas was strong. The uneven spread of unemployment across the country, then, may be an important factor in determining the dispersion of new steel investment. It may also be a vital factor in determining the possibility of closures—and this is already fore-shadowed in an observation in the First Report on Organization that

'the social problems associated with closures would be eased if units whose future may be limited were linked, so far as possible with reasonably near-by units likely to be centres of growth' (loc. cit., para. 56 (e)).

Thus despite the declared intention of the new Corporation 'to promote the investment and technological advances needed to maintain and improve the role of the public sector' in British economic growth and as a world competitor, the problem of the location of new investment remains. An all-out pursuit of this declared objective would almost certainly mean a more rapid absorption of technical advances than has happened in the past, plus a more trenchant attitude to closures, culminating in a sharp change in the labour-force level and geographical distribution.[2] In the end, however, it is likely that social and political factors will

[1] There are much larger works units in the USA, Japan and ECSC, and further mergers and joint operation schemes have recently been undertaken abroad— though usually under the surveillance of either anti-trust bodies or the ruling authority.
[2] In this connection it is interesting to note that the British Steel Corporation had already reported (by September 1968) the closure of 10 works or major departments, affecting about 2,200 men. Agreements have been entered into by the Corporation on the procedures to be followed when closures are projected: see *Annual Report*, 1967–8, pp. 30–1.

cause the objective of economic efficiency to be pursued in a more moderate fashion so that the rate of change may be no greater and perhaps even rather less than that achieved over the last few years of private enterprise.

One of the main determinants of this may be the pricing policy adopted. The pricing systems adopted by the Iron and Steel Board were explicitly formulated with the intention of stimulating efficiency. Under the system in industry-wide use over the last seven years of the Board's rule, costs were based on an ideal—the costs likely to rule in a plant of optimum size, well located and efficiently run, making use of best practice techniques whether or not they were actually in operation in British works. This system should have been effective in persuading companies of the merits of investing in lower-cost, large scale plant, but while the industry was not slow to adopt new techniques the very structure of the industry may have conditioned the Board's conception of optimum size. Also, it is not clear how far the acknowledged over-manning of the industry was taken into account in the assessment of costs in an 'efficiently-run' works. Further obscurities exist because of the lack of knowledge on the profit margin allowed by the Board, and of the effects of the 1962 change in the standard of capacity utilization on which costs were to be based.[1]

At the time of writing the Corporation's first approach to pricing policy is just becoming known and cannot be elaborated upon here.[2] Clearly the ultimate view taken about the need for a pricing system to encourage and indeed enforce greater efficiency and productivity, and the decisions taken about the appropriate level of capital utilization for normal costing, plus the imposition of a financial objective akin to that of other nationalized industries, will be among the most important factors which will determine the built-in incentive to technological change. But even when all these determinants are set out, the ultimate constraint on action is likely to be the political and social implications of an economic policy strictly applied.

The final factor which may influence the rate and consequences of technological change is the role of the trade unions in the nationalized sector. The statements of short-term and long-term objectives

[1] This standard was reduced from 90 to 84 per cent in 1962. For a fuller discussion of pricing policies, see M. Howe, 'The Iron and Steel Board and Steel Pricing, 1953–67', *Scottish Journal of Political Economy*, February 1968.
[2] This approach is described and criticized in the Report of the National Board for Prices and Incomes, on Steel Prices: *Report*, No. 111 (Cmnd. 4033, HMSO: London, 1969).

by the Corporation include reference to the need to improve productivity by achieving efficient manning standards: and improved consultation and negotiating machinery are to be developed to produce 'labour practices that will permit the highest possible productivity commensurate with fair employment practices and the social responsibilities of the Corporation'.[1] As we have seen, the trade unions organizing the steel labour force have not, in general, been responsible for any serious delay in the implementation of new techniques. A tougher attitude, both to the manning of new plant and to that on existing plant, could however result in increasing friction, even though current intentions are to improve consultative and negotiating procedures. If, as seems to be intended, union representatives are brought into consultation, at an earlier stage than under private ownership, on questions such as the introduction of major technological changes and rationalization and the creation of more efficient manning standards, it may be much more difficult to make progress to the stage at which changes are actually brought into effect. From that point of view, the rate of absorption of technological advances might, at least for the first five years or so, be slower than in the recent past. This could be damaging, since developments abroad indicate the need for fairly immediate action on this, as on a number of other problems now facing the industry.

To summarize: up to a point, the immediate future of the industry is dependent on decisions which were taken before the principal companies were nationalized. But this is not true all across the board, and there are some areas where fairly rapid decisions are required to clarify the situation, especially in the fields of investment and pricing policy and their relationship to technological change; and of manpower utilization and labour relations. The employment effects of technological change could be magnified, as compared with recent experience, if economic arguments are allowed to dominate political and social factors. But if economic arguments prevailed in this way, the manpower changes due to the movement towards more efficient manning standards of existing plant could well outweigh the effects of technological change. In the last resort, however, one must expect, as in other nationalized industries, that the social-political arguments for slower rates of economic change will be at least as significant as the economic.

[1] *Report* on Organization, para. 52 (b).

PART FOUR
THE CHEMICAL INDUSTRY

CHAPTER 8

THE STRUCTURE OF THE
CHEMICAL INDUSTRY

I. INTRODUCTION

In examining the effect of technological change on employment in the chemical industry, we shall use mainly case-study material, and this chapter indicates how the structure of output and employment in the industry have been affected by the longer-term process of change, though not only technological change. We shall first discuss generally the background of the industry so that the non-specialist reader may appreciate the nature of the industry and its products.

The most significant point is that the industry is much more complex than other manufacturing industries, and since the activities of processing, making or compounding chemicals are more or less incomprehensible to a non-scientist, many of the processes by which chemicals are manufactured are difficult to describe without using technical terms. For example, it is probably much easier to understand the process by which a motor vehicle is made than how a plastic teaspoon comes into the world. The industry's complexity lies not only in the mysterious nature of science, but also in the enormous variety of products and processes which it contains. The output of the chemical industry consists of tens of thousands of distinguishable products, and it has been estimated that about 400 new products are introduced each year. However, even the definition of a 'product' is not straightforward. One part of the industry is engaged in chemical manufacturing and uses as its raw materials elements or chemical compounds which are put through stages of processing to emerge as different chemical compounds. Other sectors of the industry simply mix or combine chemicals in such a way that a new product may contain different proportions of the ingredients without any chemical reaction or transformation being involved. It is difficult to say how many new products are chemically different and how many are simply different 'mixtures', but both are important. For example, in the large sectors producing pharmaceuticals, soap, detergent,

197

toilet preparations, etc., new products in the sense of new mixtures of existing chemicals are common. Of course, in the manufacturing process a very minor change in chemical structure means a new product has been created, and even a small change could radically alter the properties of the product and lead to new applications, or perhaps further new products based on it.

The number of products reflects the enormous variety of ways in which chemicals are used. Many of the products of the chemical industry are sold to the final consumer (e.g. soap, paint, drugs, detergents), and many others are used either in other industries as raw materials or essential additives, or in the chemical industry itself as the basis for further stages of processing, compounding or mixing. The reason for the industry's intensive consumption of its own output is that the production of many chemicals for final sale involves many stages of processing—and many intermediate products—before the final product appears. This necessary linking of processes means that physical linkages between units of equipment or complete plants are technically desirable or even essential, and hence chemical plants are often found clustered together on a single large site. The technical and economic conditions favouring the concentration of plants are often created by the nature of chemical processes which involve the output of several joint products, and if these are to be further processed and sold, there will be a complex of linkages between the basic plant and its offshoots. For example, the production of 100 tons of chemical A may necessarily involve the joint production of 60 tons of chemical B, 35 tons of chemical C, and so on often up to very large numbers of by-products. If these by-products can profitably be processed further and sold, this will improve the economics of producing chemical A, and the plant producing A may be linked physically to offshoot plants using B, C, etc., as raw materials. The best example of these technical and economic linkages is the clustering of petrochemical plants round oil refineries, by-products from which act as basic feedstock for petrochemicals of many different kinds. Similarly, ethylene is a basic 'building block' for plastics materials but a plant designed to make ethylene will also produce propylene, butylene, butadiene and petrol in varying proportions, and these normally go to be further processed in linked plants.

Of course, large concentrations of chemical plants on a particular site often contain units which have no technical relationship whatsoever, and which could just as easily have been sited in different places. Concentration of this kind may occur because chemical plants, though different, require similar conditions and facilities which are easily available in relatively few areas and there are external economies to be obtained by location in these areas. Many chemical processes are smelly, require vast quantities of cooling water, and produce large amounts of effluent which must be disposed of. A site which can fulfil these requirements—and possibly also access to good harbour facilities—will attract many plants which need not be technically linked. Also, of course, once a company has obtained a large area of land, there may be economies of construction, utilities provision, etc., which lead it to concentrate more and more plants on that site.

The complexity of products and processes in the chemical industry and the linkages between them can make the definition of the industry a matter of some difficulty, especially if one wishes to be scientifically accurate. Various definitions have been suggested to eliminate inconsistencies, and to separate the sectors which manufacture chemicals from those concerned with mixing or combining. The official definitions have been criticized in some quarters for their lack of scientific purity; S. P. Chambers pointed out that they include 'activities not normally regarded as belonging to the chemical industry, such as the manufacture of hoof meal and night lights', and that synthetic fibres production is not included, though synthetic resins and plastic rods (not fibres) are.[1] From our point of view, though, the definitional problem can be avoided, for official statistics of output and employment are based on the Standard Industrial Classification Order IV 'Chemical and Allied Industries'.[2] In fact Order IV is usually broken down into the chemical industry, comprising Minimum List Headings (MLH) 271–277, and the allied trades, MLH 261–263. These allied trades are Coke Ovens & Manufactured Fuel, Mineral Oil Refining, and Lubricating Oils & Greases. In discussion of the chemical industry, they are usually omitted and this will be the normal practice on our study. Where they are included, the terminology will then be 'Chemical & Allied Industry', and in some instances the MLH sectors will be discussed individually.

The chemical industry is very largely a twentieth-century phenomenon. Though many chemical processes have been known and understood for hundreds of years, it was only in the nineteenth

[1] S. P. Chambers, 'The British Chemical Industry', *National Provincial Bank Review*, August 1960.
[2] A full description can be found in Central Statistical Office, *Standard Industrial Classification*, (HMSO: London, 1958), pp. 11–12.

century that their industrial application began, and then only on a relatively narrow front. In 1907, the gross output of the Chemical & Allied industry was £90 million, whereas by 1963, output was some thirty times greater at about £2,700 million. It was during and after the First World War that the chemical industry experienced a rapid expansion in both its range and scale of activities, and as we shall see, growth in the recent past has also been very rapid.[1] In the earliest development of the industry during the nineteenth century, the most important activity was the production of *heavy inorganic* chemicals such as sulphuric acid, caustic soda, and other acids and alkalis. These were essential to the rapid growth of many other industries, notably textiles, papermaking and metal manufacturing, and today inorganic chemicals are part of the basic production process in almost every manufacturing industry. The *dyestuffs* sector was again a product of the nineteenth-century expansion of the textile industry, but other sectors reflect later developments. The present *fertilizer* sector, producing mainly nitrogenous fertilizers, dates from after the First World War. Previously, Britain had depended on a small output of artificial phosphatic fertilizers, and on imported natural nitrates. Only Germany had perfected an economic process for the manufacture of ammonia, a process which was developed in Britain during the 1920s and formed the basis of fertilizer production until the early 1960s. Similarly, the large-scale manufacture of *explosives* from inorganic substances was developed rapidly during the First World War.

One of the most important and most recently established sectors of the industry is the manufacture of *organic chemicals* and their derivatives. Organic chemicals, as the name suggests, were originally produced from matter derived from living organisms and especially from coal; the early dyestuffs and pharmaceutical industries were founded upon the extraction of compounds from coal. More recently, organic chemicals have been synthesized, and their production is now coming to be based on petroleum and natural gas rather than on coal. An almost infinite variety of products are based on a few organic chemicals. Most important, perhaps, are the *synthetic resins* and materials from which plastic goods are made, but organic compounds also form the basis for dyestuffs, drugs, synthetic foods and food chemicals, synthetic rubber, and petroleum derivatives of all kinds.

To summarize: the chemical industry is in several ways a very

[1] It is interesting to note that 40 per cent of ICI's home sales in 1967 were of chemicals which were unknown in 1939 (*Chemical Age*, August 3, 1968).

200

complex industry. It produces many thousands of separate products either for sale to the public or, more commonly, to act as a raw material for another process in the chemical industry or some other industry. The actual processes in the industry can be highly complex in that they stretch scientific and engineering technology to the limits but they may equally well be simple processes which have gone on unchanged for decades. In discussing the labour-force effects of technological change, we shall return to many of the points made here, notably the importance of linkages between processes, and the expansion of the production of organic and other chemicals from oil rather than coal. At the moment, however, we must examine the structure of output, employment and industrial relations in the industry.

II. OUTPUT, EMPLOYMENT AND PRODUCTIVITY

One of the most remarkable aspects of the chemical industry since the war has been the rate of growth of output. The industry grew faster than any other sector of manufacturing industry, and Table 8.1 shows the trend of output in the industry and its sectors for selected years between 1958 and 1967. Incomplete data for 1950 and 1954 are included to illustrate the experience of earlier years. Between 1958 and 1967, the output of the chemical industry grew at an average rate of 6·4 per cent per annum, twice the rate of increase shown by the Index of Industrial Production. The growth of the industry was almost as fast in the years before 1958, and from 1950–8, the average growth rate was 6·2 per cent per annum. However, the annual average rate of growth is misleading in that it suggests a continuous path of rapid growth, whereas the industry's experience was for very fast growth in some years to be partially offset by sluggish growth in others, though at no time since 1958 did the output of the industry decline from one year to the next. As one might expect from the wide usage of chemical products, the output of the industry as a whole is considerably affected by the rate of growth of industrial production in general. Thus in the two years from 1960 to 1962 when industrial production all but stagnated, the index of output of the chemical industry increased from 123 to only 131 (1958 = 100). Similarly, the industry grew considerably more slowly after 1965, as Table 8.1 shows, again largely due to the sluggish rate of growth in industrial production.

The sectors of the chemical industry showed different rates of growth of output and different degrees of fluctuation. The Plastics

TABLE 8.1

Output of the Chemical Industry 1950–67 (1958=100)

	Per cent weight[2]	1950[3]	1954	1958	1959	1960	1964	1965	1966	1967	Compound growth p.a. 1958–67
Dyestuffs	3·9	124	137	100	125	141	179	171	158	151	5·2
Fertilizers	4·2	89	85	100	108	117	131	133	129	139	3·7
Organic chemicals	11·0	—	74	100	119	147	220	232	246	262	11·3
Inorganic chemicals	10·6	—	92	100	111	122	142	150	152	159	5·3
Other chemicals; gases	19·1⎫										
Explosives	5·1⎬	—	84	100	107	116	137	141	145	150	4·6
Miscellaneous chemicals[1]	5·1⎭										
Plastic materials	7·8	32	59	100	123	147	222	241	255	278	12·0
Paint & varnish	9·3	76	84	100	110	113	122	124	125	128	2·8
Pharmaceuticals	10·8	—	74	100	113	125	165	187	206	220	9·2
Toilet preparations	3·4	73	89	100	110	122	152	154	161	169	6·0
Soap, detergents, etc.	6·4	85	93	100	103	106	110	106	107	108	0·9
Vegetable & animal oils, fats	3·2	—	103	100	106	106	106	106	102	102	0·2
CHEMICAL INDUSTRY	100·0	62	84	100	112	123	153	161	167	175	6·4
Index of Industrial Production	—	83	94	100	105	113	123	132	133	133	3·2

Notes: 1. Adhesives, polishes, insecticides, disinfectants, etc.; printing ink.
2. Per cent of net output in 1958 (excluding amount paid for services.)
3. Not based on 1958 Standard Industrial Classification.

Source: *Board of Trade Journal*, various years.

and Organic Chemicals sectors experienced average growth of 11 to 12 per cent per annum, whereas Soap, Detergents, etc., and Vegetable and Animal Oils and Fats hardly increased at all. Some of the sectors show the same pattern of variations in output as the whole industry, for example, Other Chemicals. This large miscellaneous sector, whose products are widely used in other industries, showed a check in its rate of growth of output during periods when the Index of Industrial Production was virtually static. One sector, Dyestuffs, experienced very wide variations in output, and its level of output in 1967 was well below its 1964 peak. Even the fastest-growing sectors experienced a falling-off in their rate of growth after 1965, though by other standards the rate was still rapid. The growth of the Plastics sector is closely related to that of Organic Chemicals since it is estimated that about 45 per cent of organic chemicals output goes to the production of plastics materials, though organic chemicals are also used in such rapidly expanding industries as synthetic rubber and man-made fibres. The Pharmaceuticals sector, much of whose output goes to final demand or to export, was very little affected by general economic conditions, and maintained an average growth rate of over 9 per cent throughout the period 1958–1967.

For comparative purposes it would be desirable to present employment data for the chemical industry on the same basis as the Table 8.1 output data, but two statistical problems prevent this. One is the change of Standard Industrial Classification in 1958, which changed the definition of the chemical industry itself, and reduced the scope of the Old Order IV; also the 'allied trades' (MLH 261–263) were not separately identified before 1958, so that any estimate of employment in the chemical industry proper would be imprecise. The second problem is that the Board of Trade sectoral breakdown as used in Table 8.1 does not on the whole follow the MLH scheme. Some Board of Trade sectors are equivalent or can be made equivalent to a Standard Industrial Classification MLH, but the most important Board of Trade sectors are a re-arrangement of several MLHs, and in particular of the large MLH 271, Chemicals & Dyes. No reclassification of employment data on Board of Trade lines is possible, and this is unfortunate since the classifications used in Table 8.1 are much more descriptive and useful than one based simply on the MLH system. However, in Table 8.2, which shows the number of employees in the chemical industry, the breakdown is by Minimum List Heading only.

The chemical industry is not particularly large in employment

203

TABLE 8.2

Number of Employees in the Chemical Industry 1959–67
(June each year)

	1959	1960	1961	1962	1963	1964	1965	1966	1967
Chemicals & dyes	220·8	219·9	220·0	219·6	221·7	226·0	225·0	227·3	230·8
Pharmaceuticals & toilet preparations	66·4	72·6	73·7	74·1	77·0	74·0	79·1	80·4	77·5
Explosives	37·2	33·9	32·7	30·8	28·7	26·4	25·5	30·9	27·8
Paint, printing ink	48·9	49·3	49·2	48·8	47·8	47·8	47·8	48·2	46·9
Vegetable & animal oils, fats, soap, etc.	46·4	45·4	45·2	44·1	44·1	43·2	40·2	38·6	36·3
Synthetic resins & plastics	22·6	31·1	32·9	33·9	30·7	31·0	35·5	38·6	39·1
Polishes, adhesives, etc.	11·8	14·2	14·7	13·9	14·0	14·0	15·3	14·5	14·9
CHEMICAL INDUSTRY	454·3	466·3	468·2	465·2	463·9	462·3	468·4	478·6	473·2
MLH 261–263	69·6	68·8	67·4	59·2	57·5	54·4	54·8	52·9	52·9
CHEMICAL & ALLIED	523·9	535·1	535·6	524·4	521·4	516·7	523·2	531·5	526·1

Source: *Ministry of Labour Gazette*, various years.

terms. With less than half a million employees, it accounts for only about 5½ per cent of manufacturing employment. There was a net increase of about 20,000 in chemical employment from 1959–67, though the sectoral pattern of employment varied. Some slow-growing sectors of the industry, e.g. Explosives, showed a decline in employment, but others increased. Pharmaceuticals showed the largest absolute rise, of some 11,000, while Synthetic Resins had the largest percentage increase, of about 70 per cent, between 1959 and 1967. The allied trades (MLH 261–263) suffered an all-round decline especially in Mineral Oil Refining, whose 1967 employment of 26,800 was about 12,500 less than in 1959.

In the chemical industry proper, total employment increased from 1959 to 1961, but then declined slowly till 1964. The two years 1964–6 saw a relatively large increase in employment, followed by a slight decrease in 1967. It is interesting to note that the increase from 1964–6 occurred during a period when the rate of growth of output fell away slightly from the rapid growth of the period 1962–4 (see Table 8.1), and it was out of line with forecasts of manpower requirements made in 1964. In the National Plan industrial inquiry, it was stated that 'the industry does not expect to require any increase in its total labour force during the plan period' (i.e. up to 1970).[1]

There are many possible explanations for the increase in manpower, but two are particularly relevant here. First, during 1964 and 1965 there was a widespread adoption throughout the chemical industry of the 40-hour week. The 40-hour week creates special problems for industries where continuous shift-working is common. If the normal work-week has 42 hours, four shifts each working 42 hours completely exhaust the week of 168 hours. But if each shift only works 40 hours, 8 hours every week are uncovered. There are numerous devices by which this can be overcome, such as working more overtime or continuing to work a 42-hour week with each shift receiving 8 hours' paid leave once every four weeks. The point is, though, that unless firms manage to make more intensive use of existing manpower, it may be necessary to employ some additional workers. Certain sectors in the chemical industry make great use of shift work: for example, in 1964 59 per cent of workers in Chemicals & Dyes and 54 per cent in Synthetic Resins were on shift work, the great majority of them of a three-shift continuous basis. This might explain to some extent the increase in employment in those sectors between 1964 and 1967; employment in Chemicals & Dyes

1 *National Plan* (Cmnd. 2764, HMSO: 1965), p. 146.

rose by 5,800 and in Synthetic Resins by 8,100 at a time when the rate of growth of ouput was falling, and even in 1966–7 when industry employment was falling, employment in those sectors rose.

The second reason for the increase in total employment is probably much more important. This is the failure of the industry to improve productivity by as much as had been expected. Over the whole period 1954–67, the achievement of a rapid growth of output with an almost stable labour force presupposes a fast rate of growth of productivity and in fact output per head increased by 5 per cent per annum on average from 1954–67. However Table 8.3 shows output, employment and output per head for three sub-periods 1954–60, 1960–4 and 1964–7, and illustrates how the industry's performance fell away in the period 1964–7.

TABLE 8.3

Output, Employment and Output per Head:
Average Annual Percentage Growth

	Output			Employment			Output per Head		
	1954–1960	1960–1964	1964–1967	1954–1960	1960–1964	1964–1967	1954–1960	1960–1964	1964–1967
Chemical Industry	6·0	5·5	4·5	1·7	−0·4	0·8	4·2	6·0	3·7
Chemical & Allied	5·9	5·4	—	1·7	−0·7	—	4·1	6·1	—
Mineral Oil Refining	7·3	7·5	—	2·5	−2·6	—	4·8	10·1	—
All Production Industry	3·0	3·3	2·6	0·6	0·2	−0·2	2·4	3·1	2·8

Source: 1954–64: National Plan (Cmnd. 2764, HMSO: London, 1965), Table 2.2.
1964–7: Calculated.

The rate of growth of output per annum in the chemical industry fell progressively in the three periods, and was only 4.5 per cent in 1964–7 compared with 6 per cent in 1954–60. Output per head rose fastest in the middle period, thanks to the reduction in numbers employed between 1960 and 1964, but the rate of growth of output per head fell from 6 per cent per annum in 1960–4 to 3·7 per cent in 1964–7 because of the slower growth of output and the rise in numbers employed. The rate of growth of output per head in all production industry followed the same general pattern as did the chemical industry, though the gap between them narrowed considerably in 1964–7. Evidently productivity in the chemical industry was badly affected by the slowdown in the general rate of industrial activity and by the unexpected increase in employment in the industry. In the light of Table 8.3, it is interesting to note that, with an assumed national growth rate of 3·8 per cent per annum, the

206

National Plan forecast for the chemical industry an annual average growth rate of 8 per cent in output and 8·1 per cent in output per man, with a slight decline in employment from the 1964 level.[1] Table 8.3 also gives, for comparative purposes, data for 1954–64 for the Chemical & Allied industry, which shared very much the same experience as the chemical industry, in spite of the very fast rate of growth of both output and productivity in Mineral Oil Refining, which appears to have been offset by the sluggish growth record of the other allied sectors, notably Coke Ovens.

It is difficult to obtain data on the productivity growth of most of the sectors of the chemical industry, since output and employment are classified on different bases, but it is clear that output per man in Plastics and Pharmaceuticals increased much more quickly than in the chemical industry as a whole, while the largest sector, Chemicals & Dyes, probably followed much the same trend as the industry as a whole. In general, though, the chemical industry has shown a faster increase in productivity than any other industry. In part this is a result of the process of technological change and investment in new processes, and the next chapter discusses this in more detail. Another related explanation of increased productivity is increased efficiency in the utilization of manpower. As later chapters will show, one consequence of technological change is that some chemical companies have been leaders in recent attempts to improve manning standards and their use of labour.

III. THE STRUCTURE OF THE LABOUR FORCE

Before discussing the occupational structure of the chemical industry's labour force, we may briefly mention that there tends to be a regional concentration of chemical industry employment. A comparison of the proportions of 1967 chemical industry employment and all manufacturing employment in particular regions shows that the chemical industry tended to be concentrated in the North Western region (22·7 per cent of chemical employment and 15·3 per cent of all manufacturing employment), in the Northern region (10·8 per cent and 5·3 per cent respectively) and, to a lesser extent, in the South East. These regions also specialized in particular chemical activities. The South East had a heavy concentration of employment in consumer chemicals and in mixing and activities such as Pharmaceuticals, Paint, Polishes and Adhesives. The North Western region had more than half the total employment in the

[1] *National Plan*, p. 27.

207

Soap, Detergent sector and in the Vegetable and Animal Oils and Fats sector, and it also had a high proportion of Chemicals & Dyes employment. These concentrations of employment are often quite highly localized within a region because of the tendency for chemical plants to be sited close together in favourable locations. Their existence can effect a company's policy with regard to technological change and employment, since alternative sources of employment in the chemical industry may be locally available. We shall return to this point in Chapter 10.

On the occupational structure of the chemical industry labour force, the most useful source of data is the annual survey by the DEP. These began only in 1964 so that longer-term occupational changes cannot be properly assessed, but Table 8.4 shows the

TABLE 8.4

Occupational Structure of the Chemical Industry
by Broad Occupational Group: May 1967

	All Employees	
	Number	Per cent
A. Administrative, technical and clerical workers	162,310	38·6
B. Workers in skilled occupations (apprenticeship or equivalent training)	69,440	16·5
C. Production workers with a degree of skill by training or experience (including those with between one and six months' training)	78,730	18·7
D. Other employees (warehouse, transport workers, canteen staff, labourers, others)	110,090	26·2
Total	420,570	100·0

Note: The inquiry covered a sample of firms, excluding all firms with fewer than 11 employees.

Source: Data supplied by the Department of Employment and Productivity.

number and proportions of employees in broad occupational groups at May 1967. The Table excludes the allied trades MLH 261–263, and covers about 90 per cent of employment in the chemical industry. One immediate point of interest is that administrative, technical and clerical (ATC) employees made up a very high proportion of the industry's employment, at over 38 per cent. This compares with about 24 per cent for the whole of manufacturing industry, and 31 per cent for engineering and electrical goods, the next highest industry. Being capital-intensive, the chemical industry employs relatively few production workers, so that the ratio of ATC workers to total employment is high, and because the industry is science-

208

based it employs many scientists, technologists and technicians. They accounted in 1967 for almost 25 per cent of ATC employment and 9 per cent of total employment, a proportion which only the electrical goods sector of the engineering industry could approach.

Parts B and C of Table 8.4 can be taken to refer to skilled and semi-skilled workers respectively though, as we shall see, the definition of skill is a variable one and difficult to make precise, since it can differ between industries.[1] Nevertheless a rough comparison with the relevant statistics for other industries shows that the chemical industry had, on average, a much lower proportion of workers with some degree of skill—about 34 per cent—than any other industry except Food, Drink & Tobacco. To some extent, of course, this simply reflects the higher-than-average proportion of ATC employees, but it is also affected by the large number of 'Other Employees' (Part D), most of them labourers or unclassified workers, who made up 26 per cent of the labour force. This proportion of Other Employees was exceeded by only two industrial groups, Bricks, Glass & Cement and again Food, Drink & Tobacco.

Table 8.5 shows the occupational structure of the production labour force in the chemical industry. It excludes ATC employees but includes foremen and charge hands.

The production labour force as defined here does not consist only of those employees who man or maintain chemical plants. It also includes warehouse workers, transport drivers and canteen staff, as well as the large undifferentiated groups of labourers and other employees, many of whom are employed on plants, e.g. as trade mates. The two major groups, craft workers and process workers, are most important from our point of view. About one-seventh of the production labour force was composed of skilled craft workers, almost all of whom were presumably apprentice-trained. By far the most important sub-group was maintenance fitters, millwrights and mechanics, and almost half of all craft workers belonged to this group. Next most important were other skilled engineering workers and electricians. One group not very numerous either in absolute or percentage terms in 1967 was instrument artificers, who comprised only 6·6 per cent of all craft workers. However, there is evidence that this group is becoming more important, and with the increasing number of large, complex and highly-instrumented plants, shortages of suitably qualified instrument artificers have occurred. The *National Plan* noted that

[1] See *Employment and Productivity Gazette*, January 1968, for details of the definition.

TABLE 8.5

Occupational Distribution of the Production Labour Force
in the Chemical Industry: May 1967

	Number	Per cent of Manual Labour Force	Per cent of Sub-group
CRAFT WORKERS			
Fitters, millwrights, etc.	17,830	6·9	48·6
Electricians	4,930	1·9	13·4
Instrument Artificers	2,470	1·0	6·7
Other skilled engineering	5,570	2·2	15·2
Bricklayers	850	0·3	2·3
Carpenters/joiners	1,850	0·7	5·0
Other skilled building	2,060	0·8	5·6
Other	1,120	0·4	3·1
Total Craft	36,680	14·2	100·0
PROCESS WORKERS			
Skilled	4,090	1·6	4·0
Highly trained	19,920	7·7	19·4
Semi-skilled	78,730	30·5	76·6
Total Process	102,740	39·8	100·0
FOREMEN/CHARGE HANDS	8,750	3·4	—
OTHER EMPLOYEES (Including Labourers, Warehouse, Transport Workers, etc.)	110,090	42·6	—
Total Production Labour Force	258,260	100·0	—

Note: For fuller details of process worker and other employee classifications see Table 8.4 and the text.

Source: Data supplied by the Department of Employment and Productivity.

such shortages existed in 1965,[1] and most company forecasts of future manpower trends in the chemical industry foresee continuing shortages for several years to come.

Process workers accounted in 1967 for about two-fifths of the production labour force but the classification of process workers in Table 8.5 is not very precise. 'Skilled' here means an occupation to which entry involves apprenticeship or an equivalent training; 'highly trained' describes occupations where skill is acquired by several years' experience or a minimum of six months' training,

[1] *National Plan* (HMSO: London, 1965), p. II.63.

while 'semi-skilled' describes workers who have 'acquired a degree of skill by experience and/or some training or experience before becoming reasonably proficient'.[1] There are obviously difficulties in this classification, since for the same job one employee may undergo a different training period from another. For example, one company may believe that a particular job requires a training period of eight months, while in another an employee doing the same job may receive less than six months' training and yet be regarded as equally proficient: he would be regarded as semi-skilled while in the former case the worker would be highly trained. Obviously too, the length of the training period may be just as much a function of the aptitude or ability of the trainee as of the complexity of the work or the skill necessary to do it.

The category of skilled process operators is rather more precise. The requirement of apprentice training or its equivalent suggests that many of these employees have undergone formal training in a company training programme, or have become Qualified Chemical Operators under the four-year course administered by a joint committee of the Chemical & Allied Industries Joint Industrial Council (see below). Only 4·0 per cent of process workers (1·6 per cent of the production labour force) fell into this category in 1967. Production workers who are not skilled in this sense, but are highly trained, accounted for about 19 per cent of the process labour force, but more than three-quarters of process workers were, by the DEP definition, semi-skilled. The requirement of between one and six months' training is significant, for it indicates the relative simplicity of most of the actual process operations in an industry which is scientifically and technologically most advanced. The scope for skill of a manual or mechanical type, such as is traditionally exercised by craftsmen, would appear to be very small, and the fact that over 70 per cent of the process labour force consists of semi-skilled and unskilled workers is a reflection of this.

IV. THE INDUSTRIAL RELATIONS BACKGROUND

In view of the high proportion of white-collar workers in the industry, it is perhaps reasonable to expect that trade unionism in the chemical industry would not be particularly strong, but it is nonetheless a little surprising to find how few of the employees in the industry are members of a trade union. In 1964 it was estimated that

[1] Ministry of Labour description quoted from *Ministry of Labour Gazette*, January 1967, p. 17.

there were 186,000 trade union members out of a total employment in the Chemical & Allied industry of 515,000, giving a proportion unionized of just over 36 per cent.[1] This compared with 41 per cent for the whole of industry and commerce, and with very much higher figures for other manufacturing industries.

Unfortunately it is impossible to estimate trade union membership in the chemical industry for any year later than 1964, because of the structure of membership. Most process workers in the industry are members of either the Transport and General Workers' Union or the National Union of General and Municipal Workers, and the DEP, in its annual survey of trade union membership, is unable to allocate the members of these general labour organizations among the appropriate industry groups. Other trade unions have some members among process workers, notably the Union of Shop, Distributive and Allied Workers, whose members again are not allocated to the chemical industry in DEP data, and the Chemical Workers' Union. The CWU has a long and bitter history of struggle with the larger general unions over recognition and negotiating rights in the chemical industry.[2] It was expelled from the TUC in 1924 for 'poaching' members from other unions and was not re-admitted until 1943, but since that date it has failed to secure substantial recognition in the negotiating machinery of the chemical industry, nor has it made much progress in recruitment: the union has only about 16,000 members. In 1966, the NUGMW proposed that the CWU should merge with it, an amalgamation which would have resulted in a larger and more efficient chemical workers' section within the NUGMW.[3] The executive of the CWU approved the amalgamation terms, but at the union conference the members decided not to accept them, preferring instead to continue as an independent trade union.

The maintenance workers in the chemical industry are organized by the union appropriate to the craft in which they work. On the basis of the occupational distribution of the labour force, it appears that the most important craft unions would be the Amalgamated Engineers and Foundry Workers and the Electrical Trades Union, with fewer members from the building trade unions and the unions

[1] *Selected Written Evidence Submitted to the Royal Commission on Trade Unions* (HMSO: London, 1968), Fourth Supplementary Memorandum by the Ministry of Labour, p. 23.
[2] For an historical analysis, see S. W. Lerner, *Breakaway Unions and the Small Trade Union* (Allen & Unwin, London, 1961), Chapter 1.
[3] For details, see National Union of General and Municipal Workers, *Written Evidence to the Royal Commission on Trade Unions.*

for the 'black trades', such as boilermakers, plumbers, and sheet-metal workers. There is little official information on the degree of unionization of white-collar workers, but since they are so numerous in the chemical industry, one might expect a reasonably high level of unionization. The most recent study showed exactly the opposite. This analysed the membership of the major white-collar unions (accounting for 95 per cent of manufacturing white-collar trade unionists), and found that only about 3·7 per cent of all white-collar workers in chemicals were union members, mainly in the Association of Scientific Workers, the Association of Supervisory Staffs, Executives and Technicians, and the Transport and General Workers' Union.[1] The author also concluded that, in general, white-collar unions faced a problem in achieving recognition by chemical industry employers.[2]

The industrial relations machinery on the employers' side is easier to describe. For the chemical industry proper (i.e. the manufacture of chemicals, rather than the mixing and compounding activities), the relevant body is the Chemical Industries' Association, formed in 1966 by the amalgamation of the trade association and the employers' association for the industry. The CIA's member firms employ about 250,000 workers, about 50 per cent of the whole industry's employment. Basic negotiations on wages and conditions of employment of process workers are, for most firms, carried on by representatives of the CIA and the unions on one of several negotiating bodies.[3] Most important of these is the Chemical & Allied Industries Joint Industrial Council, which has three separate subsidiary JIC's, for the Heavy Chemicals, Plastics, and Fertilizer sectors respectively. Another joint body on which the CIA and the general unions are represented is the Drug and Fine Chemical Joint Conference, and there are separate JIC's for workers in gelatine and glue, in paint, varnish and lacquer, and in soap, candles and edible fats. For maintenance workers, negotiations proceed between CIA and the craft unions concerned. However, not all companies are members of these negotiating procedures, and foremost among these is ICI which, though a member of the CIA, negotiates its own national agreements with the trade unions. Some other companies

---

[1] G. S. Bain, *Trade Union Growth and Recognition* (Royal Commission on Trade Unions, Research Paper No. 6: HMSO: London, 1967), Tables 16 and 14.
[2] Ibid., Table 18, p. 73.
[3] A description of this machinery can be found in Ministry of Labour *Industrial Relations Handbook* (HMSO: London, 1961), pp. 46–9. The Association of Chemical and Allied Employers referred to therein has been superseded by the Chemical Industries Association.

similarly set wages and conditions of employment by bargaining with both union national officers and local representatives.

What has this negotiating machinery accomplished for workers in the chemical industry, in terms of improvements in wages, and in non-wage conditions of employment? Table 8.6 shows the level of

TABLE 8.6

Hourly and Weekly Earnings in the Chemical Industry:
Male Manual Workers: October 1968

	Average Weekly Earnings Shillings	April 1960 = 100	Average Hourly Earnings Pence	April 1960 = 100
Coke ovens, etc.	428/0	156·4	112·6	153·6
Mineral Oil Refining	533/7	178·5	151·0	198·9
Lubricating Oils, etc.	438/9	163·7	111·8	175·5
Chemicals & dyes	484/8	165·1	126·2	169·4
Pharmaceuticals, etc.	429/1	169·1	112·2	172·1
Explosives, etc.	435/1	157·8	113·7	163·8
Paint, etc.	425/3	162·1	113·9	174·4
Vegetable/animal oils, fats, etc.	510/11	170·7	124·6	170·7
Synthetic resins	472/-	156·3	125·3	161·9
Polishes, etc.	460/1	168·0	110·4	170·9
CHEMICAL & ALLIED	472/11	165·0	123·6	170·7
ALL MANUFACTURING	472/4	159·4	123·8	165·1

Source: Department of Employment and Productivity, *Statistics on Incomes, Prices, Employment & Production.*

and increase in the earnings of male workers. A comparison between All Manufacturing and the seven sectors making up the chemical industry, shows that for weekly earnings, five out of the seven were lower than the manufacturing figure, while for the hourly earnings four out of seven were lower, though earnings in Chemicals & Dyes, by far the largest sector, were higher than the All Manufacturing figure. On this data, it seems that the chemical industry is not a particularly high-wage industry, but one has to bear in mind that the chemical industry has relatively few skilled workers. Taking into account the low proportion of skilled workers—who are likely to be high wage workers—one might conclude that the industry has a high level of earnings for the type of labour it employs. In addition, hourly and weekly earnings have shown a relatively rapid rate of increase since 1960; only two SIC Orders have shown a faster increase in weekly earnings than Chemical & Allied Industries, and four have increased hourly earnings more quickly. But a

214

comparison of the chemical industry proper would require to omit the allied trades, especially Mineral Oil Refining which showed a much faster increase both in weekly and hourly earnings than any other chemical industry MLH, and though earnings increases in most of the sectors of the chemical industry proper were above the average for all manufacturing, a number of other industries and sectors showed a better record.

There are two sources of data on non-wage conditions of employment in the chemical and allied industry, and both show the industry to be one of the leaders in this field. One was an inquiry into fringe benefits in the year 1960, carried out by the University of Glasgow. This covered companies employing about one-third of the manual workers in the industry, mainly the large companies, and found that 84 per cent of companies had sick pay and pension schemes, 94 per cent had works canteens, and a relatively high proportion spent money on various other fringe benefits.[1] The amount spent, both in money terms and as a percentage of payroll, was also the highest of any industry group, except that Food, Drink & Tobacco had a higher expenditure relative to payroll. A more recent source is the Ministry of Labour's survey of labour costs in manufacturing industry in 1964, and this confirms the impression of the chemical and allied industry as a leader in welfare. Both in money terms and relative to wages, the industry had much the highest expenditure on private social welfare payments (including pensions, redundancy, sickness, accident, etc.) and on subsidized services of various kinds.[2] An analysis of wages and salaries shows that the industry spent more as a percentage of wages and salaries on sickness benefit and on holidays than did any other industry.[3]

In dealing with redundancy, many firms in the chemical industry have also been ahead of normal industry practice. The Chemical & Allied Industries JIC had by 1961 a recommended procedure for dealing with redundancy. This simply set out the kind of principles which companies ought to follow when faced with a redundancy, and it had no binding force on member firms. However, some of the large companies in the industry had gone beyond this and had formulated redundancy policies and schemes, either as a general

[1] G. L. Reid and D. J. Robertson (eds.), *Fringe Benefits, Labour Costs and Social Security* (Allen & Unwin, London, 1965), Table 25, p. 73.
[2] 'Labour Costs in Britain in 1964: Manufacturing Industries', *Ministry of Labour Gazette*, December 1966, Table 4.
[3] Ibid., Table 5. This tends to confirm the impression that the chemical industry has more liberal conditions for extra holidays. See Department of Employment and Productivity *Time Rates of Wages and Hours of Work*.

framework for dealing with any redundancy which should arise, or to cope with particular situations. An outstanding example of the former approach is the ICI Protection of Employment scheme, which basically aimed at avoiding redundancy, but also laid down principles for dealing with it should it ever arise. In general, it seems to be normal for these redundancy schemes, and the pension and welfare schemes, to be proposed by the company and agreed by the trade unions rather than for them to be negotiated on a formal basis.

Finally, we must note that these wage and employment conditions have been set in a framework of union-management co-operation. Since the war, the industrial relations record of the industry has been very good, and the number of days lost through strikes has characteristically been well below the national average and very much lower than most major manufacturing industries.[1] This is not to say that the unions have been ready to acquiesce in every proposal of management, but in general there has been a readiness to see the importance of changes which facilitate the continued expansion of the industry, and the irrelevance of maintaining restrictions for their own sake. At any time, managements of the large firms have generally been willing to see the trade unions' point of view, and the two sides have been able to discuss the issues in a constructive fashion.[2] It is significant that many of the original comprehensive productivity agreements occurred in the chemical and allied industry. A point which is closely related to this is the degree to which the leading firms in the chemical industry have regarded themselves as 'good employers'. The discussion of the non-wage conditions of employment showed how prevalent pension schemes, sick pay plans, etc., were in the industry and the proposals of productivity bargaining are again a function of progressive management. In this respect, it is noteworthy that the Chemical Industries' Association and the craft and general workers' unions have signed an agreement designed to facilitate productivity bargaining in the industry. As well as laying down the basis on which productivity bargaining should be carried on, the agreement established a joint standing committee to set up an advisory service to assist management and unions in developing productivity agreements.[3]

[1] See for example, *Ministry of Labour Written Evidence to Royal Commission*, Appendix XVIII, pp. 72–3, or *Ministry of Labour Gazette*, May 1967, pp. 384–5.
[2] Unfortunately this is not always so. During 1966–7, ICI failed to get the local officials of the AEU and other unions at particular sites to agree even to discuss a proposed productivity agreement, though eventually trials of the agreement were agreed at different sites.
[3] For details, see *Chemical Age*, December 16, 1967.

216

The relevance of this to our study of technological change is that given this legacy of union-management co-operation or at least the willingness to talk to one another, one would expect attempts to be made to ease labour-force adjustment to change as far as possible without either side striking an inflexible attitude. The existence of pension and redundancy payment schemes will automatically make the use of certain adjustment devices easier, as we shall see in later chapters, while a declared policy on the avoidance of redundancy and the approach to any inevitable shedding of manpower will enable management and unions to talk in an atmosphere of mutual confidence. In discussing the labour force adjustment to change, we must judge whether the industrial relations background of the industry did facilitate the process, or whether unexpected strains developed.

CHAPTER 9

TECHNOLOGICAL CHANGE IN THE CHEMICAL INDUSTRY

Though economists have found it difficult to define and measure precisely the contribution of technological change to economic growth, in the context of our study it is necessary to have some conception of the rate at which technological change is advancing. Chapter 1 adopted a very broad definition of technological change, as being changes in the combination of capital and labour used to produce a given output, and the best measure of the way in which this combination is changing so as to make the best use of resources, is simply the rate of increase of output per man or man-hour. Under this definition, the chemical industry as a whole has experienced a rapid pace of technological change, and indeed on most counts the industry can be considered technologically advanced, since it is capital-intensive, science-based, makes extensive use of technical personnel, is dependent on continuous research and development, and utilizes technically advanced engineering techniques and methods. Nonetheless, the industry is an aggregate of many disparate sectors utilizing a large number of process types, and many of these processes have remained unchanged for decades because of technical or economic factors favouring their retention. Also some sectors have shown a slow growth in productivity over quite long periods. It is therefore inaccurate to talk of the whole chemical industry as being subject to a rapid rate of technological change, although the present discussion will for the moment continue to deal with the industry in the aggregate.

Chapter 8 showed that the chemical industry had experienced a very rapid growth of output with little change in the numbers employed, and Chapter 10 will discuss in more detail the effect on the industry labour force and the devices used by companies in adjusting their labour force in a situation where technological changes are occurring. This chapter provides a link between the economic performance of the industry and the labour-force effects flowing

218

from it, shows how and why productivity changes occurred which allowed the industry to expand so quickly with little increase in employment, and illustrates in broad terms the types of technological change which have occurred. Section I discusses for the chemical industry potential sources of improvement in technological knowledge and the types of change to which they may give rise. Section II deals with the application of new techniques and the relationship of demand, costs and output in certain sectors, in particular the importance of economies of scale in the fast-growing sectors. Finally, Section III presents data on investment in the chemical industry, and analyses some of the problems raised in recent years by the rapid pace of technical advance and the simultaneous implementation of new techniques by many firms.

I. THE IMPROVEMENT OF TECHNICAL KNOWLEDGE

(i) *Introduction*

The chemical industry provides a good example of the various ways in which research and development (R & D) can lead to improvements in the stock of knowledge. Most companies in the chemical industry carry out some basic or fundamental research which may lead to a better understanding of the basic chemistry involved, but which has no immediate commercial application nor possibly even one in prospect. Fundamental research is not completely aimless, since it will obviously be concentrated on areas where there is some probability of a successful outcome, but the company may feel that research is worth while pursuing even where the probability of commercial application is small, because the returns from a successful outcome would be high. A second type of research which often derives from fundamental research is applied research, where efforts are directed along a particular line of inquiry from which specific and commercially viable results are expected. For example, an applied research project might attempt to synthesize a new compound which is expected to have properties making it more efficient than existing products.

Much more important, though, is the process of development which in the chemical industry commonly involves the translation of successful laboratory research to the scale of an industrial process. Problems which were easy to solve in the carefully controlled conditions of the laboratory may take on a new complexity for the chemical engineer. He has to work on a scale several hundred or thousand times larger, and this will raise new problems which did not present themselves in the laboratory, not necessarily problems of making the chemical reaction work, but rather of containing it and controlling it satisfactorily. Development of a process usually involves the construction of a pilot plant which is thoroughly tested to show up the effect of temperatures, pressures and flow of product through the plant. It will also be necessary to check that the instrumentation of the process is sufficiently sensitive to allow proper control, and that the machinery is sufficiently reliable to guarantee safe and efficient operation. Most important, the laboratory experiment cannot guarantee that the industrial application will give a product which is competitive or a process which is more economic than existing manufacturing processes.

There are therefore two areas of interest in the field of research and development, the fundamental and applied chemical research into new products and new applications of existing products, and the industrial and chemical engineering research into methods of production of new products or more efficient methods of production using substantially unchanged techniques. Both these types of research and development are carried out by chemical companies, but chemical engineering companies also carry out research into new techniques, and provide operating chemical companies with the opportunity to install new processes or improvements to existing processes. Often a chemical producer and a chemical engineering company combine to develop a new industrial process, and some very large chemical companies possess the resources themselves to develop the industrial application of a new process. The contribution of the chemical engineering industry to the development of new techniques of the chemical manufacturing industry is very considerable, but we shall largely confine our attention to the chemical industry. Even in this context, it is difficult to define and measure accurately additions to the stock of knowledge both of chemical and industrial processes, but two rough indicators exist, namely the amount spent on research and development, and the number of patents which are taken out to protect new processes and products.

(ii) *Research and Development Expenditure*
The Ministry of Technology has carried out surveys of R & D expenditure in manufacturing industry covering the years 1961–2, 1964–5 and 1966–7. Table 9.1 gives details. There was a considerable rise in expenditure in R & D in the chemical industry, and the chemical industry's share of all manufacturing R & D also increased between 1961–2 and 1964–5, though it was almost the same in 1966–7.

TABLE 9.1

Research and Development Expenditure within the Unit:
Private Industry, 1961–2, 1964–5 and 1966–7

	1961–2	1964–5	1966–7
All manufacturing (£m)	378·0	445·3	554·5
Chemical industry (£m)	39·7	55·1	68·3
of which			
Pharmaceuticals	—	—	17·0
Plastics	—	—	12·5
Chemical products	—	—	38·8
Chemical industry as % of all manufacturing	10·5	12·4	12·3

Note: Chemical industry excludes chemical engineering and the allied sectors: petroleum products spent £10·3 m. in 1966–7.

Source: Ministry of Technology.

Another useful measure is R & D expenditure expressed as a percentage of total sales, and in 1966–7 R & D expenditure was just over 2 per cent of sales. Relative to other manufacturing industries, the chemical industry is heavily research-based because of the scientific nature of the industry, and its relatively recent growth and development means that the frontiers of knowledge about products and processes are continuously being extended. The importance of research to the industry emerges even more strongly if we consider only company-financed research. Of total R & D expenditure of the chemical industry in 1966–7, over 90 per cent was company-financed, while in the other sectors where total spending on R & D was high, the electrical and aircraft industries, more than half the expenditure was financed by the Government. The chemical industry was responsible for 18·6 per cent of all company-financed expenditure, second only to the electrical and electronic engineering industry.

Among companies in the chemical industry ICI is by far the largest spender on R & D. This is not surprising in view of the size of the company, which accounts for about 20 per cent of chemical industry sales and employs well over 100,000 employees. In 1961, ICI and its British subsidiaries spent £15 million (with an additional £3 million on technical services for home and overseas companies), and in 1968 the company spent £29·3 million on R & D, though an unstated amount of this went on technical services.[1] A comparison with the data in Table 9.1 suggests that ICI accounts for at least 40 per cent of industry R & D. ICI's R & D expenditure amounts to over 4 per cent

[1] ICI *Annual Reports.*

of sales, so that the company spends twice as much on R & D, as a percentage of sales, as the average for the chemical industry. This is explained by its strong representation in the fast-growing research-based sectors such as plastics, organic chemicals, and pharmaceuticals, and by the fact that ICI itself does a good deal of research into the development and industrial application of new processes.

Statistics of R & D expenditure can only be regarded as indicative of one force making for an increase in technical knowledge, and differences or ambiguities in definition and coverage often make it difficult to say how far a particular type of expenditure really indicates an improvement in technology or represents an addition to knowledge which will allow more efficient production. It is from 'development' research that new techniques will spring for it is at this stage that the industrial application of new processes is devised, though applied research may yield discoveries which are quickly and profitably introduced with little development cost, and fundamental research too is obviously essential to the long-term progress of the industry. Even when this distinction is made, one has to look closely at the various types of R & D. In some sectors of the industry, applied and development expenditure involve the improvement of existing products. There is no change in the manufacturing process, but rather an alteration in the proportions of ingredients, and a marginal improvement in the performance of the final product. This kind of development activity is common in sectors whose output goes to the final consumer—e.g. soap and detergents, paints, toilet preparations—and in sectors such as dyestuffs where mixing and compounding are an important part of the manufacturing process.

Unfortunately, the data on R & D do not permit very fine distinctions between types and functions of expenditure, but Table 9.2 shows a breakdown of R & D current expenditure into basic, applied and development expenditure. The table shows that about 53 per cent of chemical industry R & D expenditure went on development. This was much lower than in all manufacturing, where more than three-quarters of R & D was on development. The very high proportion of chemical industry R & D which was spent on basic and applied research emphasizes the research-based nature of the industry, but the data are not comprehensive for there are other sources of R & D which might add to the stock of knowledge. For example, the proportion of research done by agencies outside the chemical industry might increase. Considerable advances in knowledge may spring from independent or Government research

TABLE 9.2

Current Expenditure on R & D by Private Industry (excluding Public Corporations and Research Associations): 1966–7

		Basic	Applied	Development	Total
All manufacturing	£m	17·9	93·5	391·1	502·5
	%	3·6	18·6	77·8	100
Chemical industry	£m	4·0	23·9	28·7	56·6
	%	5·3	42·2	52·5	100
of which					
Pharmaceuticals	£m	0·6	8·3	5·8	14·7
	%	4·1	56·5	39·4	100
Plastics	£m	0·5	4·2	4·7	9·4
	%	5·3	44·7	50·0	100
Chemical products	£m	2·9	11·4	18·2	32·5
	%	8·9	35·1	56·0	100

Source: Ministry of Technology.

institutions which are not financed by companies, and not all such expenditure would be included in the tables. Some outside research may be financed by companies, but a 1961 report found that expenditure on external research was only about 3 per cent of internal research expenditure and the Ministry of Technology survey of 1966–7 found that current expenditure of research associations, most of it basic and applied research, was less than 2 per cent of total spending.

We have already pointed out that it is development, rather than research as such, which will tend to lead to new technology and there are other sources of technological knowledge which do not appear in the R & D statistics. Licensing agreements, important in the international chemical industry, can add to a country's knowledge of new techniques without direct or indirect research expenditures. British subsidiaries of overseas parent companies may draw directly on the parent's research knowledge without conducting their own research, or British companies may use processes developed by foreign companies in return for an appropriate royalty.[1] Also, the chemical engineering industry is responsible for many of the additions to the stock of process knowledge of the chemical industry, and the amount which chemical engineering companies (many of them international) spend on development of new processes cannot

[1] ICI is the most important licensee in the British chemical industry and in 1965 it earned some £16 million from home and overseas royalties. The UK industry's export trade in licensing processes amounted to £17 million. See 'Chemicals Survey', *Financial Times*, July 29, 1968. For a general consideration of the importance of licensing, see 'Chemical Process Plant: Innovation and the World Market', *National Institute Economic Review*, August 1968.

be satisfactorily estimated. However, the National Institute study commented that

'the contracting firms in every country rely to a considerable extent on the chemical industry for new processes, although they possess the technical capacity to modify and improve these processes and can generate a few of their own.'[1]

The study found that the contractors and component makers were generally not research-intensive and had small research and development departments, though they contributed much to smaller developments of existing processes.

Of course, fluctuations in the amount of R & D expenditure or its relationship with sales cannot be taken to indicate that more or less knowledge is being acquired. Companies may regard R & D expenditure as a fixed cost which is unrelated to sales, and variations in R & D may affect those parts of it which are least relevant to applied knowledge in the short-term. During a cut-back in expenditure, it is common for fundamental research to be the first casualty, and the immediate effects of this in terms of application of techniques is likely to be limited.

(iii) *Patents*

The number of patents taken out on chemical processes gives some idea of the speed at which inventions are being produced, but again the correlation is only approximate. Patents can obviously cover the results of fundamental research which have no possibility of immediate application, or laboratory processes whose development on an industrial basis may take much further research. Again, a large number of interlocking patents can be taken out to protect one particular process against the research efforts of competitors. More important, perhaps, is the general point that a few patents may cover outstandingly significant advances in technology, while most others simply occur in the natural process of defending lines of research.

From 1963 to 1967 the number of patents registered in respect of chemical products and processes rose from 5,825 to 8,864, and in 1967 this represented 20 per cent of all patents.[2] About 40 per cent of all chemical patents were in the field of organic chemicals, one of the fastest expanding sectors strongly based on research, and about 30 per cent in polymer chemistry on which the plastics industry is

[1] 'Chemical Process Plant', op. cit., p. 39.
[2] *Report of the Comptroller-General of Patents, Designs and Trade Marks* (HMSO: 1968).

based. In an analysis of patent statistics in the plastics industry during the period 1954–62, Freeman found that from 1954–8 the 30 leading firms took out 3,119 patents, while from 1959–62 the total was 4,348, the annual average rising from 624 to 870.[1] The leading firm in both these periods was ICI, which took out 299 patents in the earlier period, and 485 in the later. The majority of the other 30 firms were German or American, presumably protecting the results of their home research against British imitation. In fact, of course, processes patented by these firms might never be introduced in Britain, unless under licence or by their British subsidiaries, and the extent to which the number of patents actually reflects the state of applicable knowledge is thereby further qualified. The later National Institute study found that chemical firms took out far more patents than chemical contractors, 'reflecting the much greater contribution of chemical firms to applied research and development'.[2]

The statistics of R & D expenditure and patents do not enable us to estimate the rate of increase of the stock of knowledge, but they do indicate how the chemical industry in general—and parts of the industry in particular—is much more dependent on research and innovation than most other industries. This is obvious from the disproportionate share of R & D expenditure and patent applications for which the industry is responsible: the industry accounts for less than 10 per cent of productive industry net output, but over 20 per cent of industrial R & D expenditure and patent acceptances. All of this simply confirms the common-sense view that potential technological progress is a powerful factor underlying the growth of the industry. We must now consider what kinds of technological change might be introduced as a result of technological progress.

(iv) *Types of Technological Change*
In the steel and printing industries, the main technological developments can be listed and described so that their relevance is obvious to the layman and the extent of the change understandable. For example, in printing the distinction between film-setting and hot-metal setting is easy to explain and the technological advance which film-setting represents can easily be comprehended, even if the details of the technique cannot. Unfortunately, the complexity of the technology involved in chemical processes makes many of the most important developments intelligible only to the specialist.

[1] C. Freeman, 'The Plastics Industry: a Comparative Study of Research and Innovation', *National Institute Economic Review*, November 1963.
[2] 'Chemical Process Plant', op. cit., p. 44.

H

Some differences can be appreciated, as in the physical appearance of an old coal-based plant and a new oil-based plant making the same product, but an explanation of *how* the plants differ could only be given by and understood by a chemist or engineer. It is therefore difficult to deal in detail with particular technological changes, and instead we must attempt to classify them very broadly according to whether the change is a major one or a minor one.

Although there is no fixed borderline between 'large' and 'small' technological changes, the difference between them is clear enough at the opposite ends of the spectrum. At one end, there is the fundamental change of technology involving different chemistry, a different type of plant, and an entirely different manufacturing process. Many examples of this type of change exist. For example, the original coal-based process for making ammonia involved passing streams of air and steam over coke at a very high temperature, and the resulting streams of hydrogen and nitrogen were synthesized to give ammonia. The modern petrochemical-type process uses light hydrocarbons (naphtha) reformed with steam at pressure as the source of the hydrogen, which is combined with the nitrogen of the air to give ammonia. At the other end of the spectrum is the relatively minor change in the production chain, involving for example the development or refinement of an existing technique at one stage of production, perhaps by such simple means as changing temperature or pressure, improving the machinery used, introducing a different raw material, and so on. There are also various other kinds of small adjustment to the production process, which would not be considered technological change in the engineering sense, e.g. a reduction of labour input through work study, giving better use of existing capital equipment. Since such adjustments are parts of every company's desire for greater efficiency, they tend to be frequent and cumulative, but they are also difficult to identify individually because their administration tends to be decentralized and not documented.

In the process of technological change as we have defined it, there are two areas where research and development have been particularly important in leading to the implementation of process improvements. First, because chemical production is often a chain of linked processes with many stages of production and several pieces of machinery integrated into a continuous process within each stage, there is likely to be differences in technology between the stages. Unless technical knowledge is improving at equal speed for all sub-processes, which is unlikely in theory and does not seem to happen

226

in practice, the efficiency of the process as a whole will be limited by the level which the technology of the least efficient sub-process has reached. Hence the implementation of new techniques often means the introduction of improvements to the 'bottleneck' production stage, since if this is removed it will allow the more advanced sub-processes to be efficiently utilized. Secondly, a trend which is apparent in the implementation of technological change in the chemical industry is the increasing size in output terms of production units. It is rare for a new plant to have smaller output capacity than the plant it replaces, for there are often substantial economies of scale in chemical processes and plants. Obviously, the introduction of new larger plants depends on a favourable combination of economic and technological circumstances, but in some sectors of the industry conditions have allowed such plants to be brought into operation. The next Section discusses in more detail how these conditions have come about.

II. THE APPLICATION OF NEW TECHNOLOGY

Underlying the decision to introduce a new technique is the calculation of its cost and the returns expected from it, and here the main factors are the trend of demand and price of the product, the capital and operating cost of the equipment, and the rate of return required on the investment. The question whether to replace an existing technique by a more efficient one is rather complicated, as we shall see, but first we may consider the trend of demand, and the importance of scale in cost of production.

(i) *The Growth of Demand*

The previous chapter showed that the output of the chemical industry and most of its sectors had increased more quickly than industrial production generally. For some sectors, rapid growth of demand and output is dependent on the growth of a single industry which is a preponderant customer. For example, the textile industry and the building industry are by far the largest consumers of dye-stuffs and paint respectively, and the growth of these industries is the major influence on the demand for these chemicals. The demand for other chemicals sold mainly to the final consumer is strongly influenced by increases in real income. The income-elasticity of demand is high for such products as toilet preparations, soap, some pharmaceutical products, and for a variety of goods using chemical raw materials or intermediate products. More

227

generally, many basic chemicals or specialized industrial chemicals are dependent on the growth of industry as a whole, because of their wide usage, and the importance of intra-industry usage of chemicals means that rapid growth of certain sectors of the industry tends to stimulate the demand for the products of other sectors.

The process of technological change in other industries tends to produce new uses for chemicals, and the increasing complexity of industrial processes has led to a rise in demand for chemicals, both for new products and larger quantities of existing products. Some of these industrial chemicals are basic raw materials, the demand for which is inelastic, in the sense that the amount used depends on the output of the final product and not on the price of the chemical. Also some chemicals have no substitute nor can they easily substitute for other products. But underlying the upward trend of demand for chemicals is the decline in the price of many chemical products relative to competitor products. Table 9.3 gives

TABLE 9.3

Index Numbers of Wholesale Prices 1968 (1954 = 100)

Basic Materials and Fuel		Prices of Various Commodities	
Manufacturing industry	110·1	General chemicals	104·2
Chemical and allied	117·0	Pharmaceuticals	103·1
		Toilet preparations	149·4
Price of Output		Paint	116·7
Manufacturing industry	134·4	Soap	154·1
Chemical and allied	109·6	Detergents	103·7
		Synthetic resins and	
		plastics materials	73·8

Source: *Annual Abstract of Statistics*, 1968.

the movement of wholesale prices, and of the prices of inputs, for the chemical and allied industry. Although the prices of inputs (basic materials and fuel) rose more quickly than those for all manufacturing, and so since 1960 have average earnings in chemicals, wholesale prices have risen by only 9·6 per cent since 1954, compared with 34·4 per cent for all manufactured products. The table shows too the trend of wholesale prices in different sectors. The prices of toilet preparations and soap rose by much more than the average of all wholesale prices, but in the large General Chemicals sector which contains most of the chemicals used as industrial raw materials, wholesale prices rose only slightly, and they fell by some 26 per cent in Synthetic Resins & Plastics Materials.

Another source of demand for chemicals, and one which has been influenced by the price stability of the industry, has been the export market. Chemical exports by value increased by about 83 per cent between 1958 and 1967, compared with about 58 per cent for all manufactured exports, and chemical exports increased relative to the output of the industry as a whole. Thus, even though Britain's share of world trade in chemicals has been declining and the balance of trade in chemicals has not been consistently favourable in all sectors, the export sector provided a ready market, especially for plastics and artificial resins, exports of which grew particularly rapidly.

(ii) *The Economics of Large-scale Production*
The growth in demand for chemicals has in part been stimulated by the relative decline in their price. Obviously any technological change which replaces an older process will only be implemented if it can show a lower cost per unit, but technological change has allowed cost and price reductions both through a pure 'technology' effect and through making possible larger and larger plants with lower unit costs. An example of the former effect can be seen from a comparison of the steam-reforming process for the manufacture of ammonia with the former coal-based process. The production cost per ton was reduced to one-quarter of what it had been, and the new process used one-fifth the amount of land, one-sixth the amount of electricity, and 0·3 man-hours per ton compared with 11·6 man-hours per ton on the old plant.[1] This spectacular reduction in unit cost was not the result of economies of scale—the output of the two processes was not very different—but of the change in technology. To estimate the contribution of economies of large-scale production in any real-life situation is rather more difficult.

Economies of scale exist where increases in capacity of plants producing a standard product result in a lower average cost per unit. One should also assume similar states of technical knowledge, and this assumption is difficult to maintain in any empirical investigation of economies of scale, for it is frequently the advance of technical knowledge which allows larger plants to be built, even though the basic technology is still the same. There are further complications in determining the exact relationship between the scale and composition of output and average unit costs,[2] but the fundamental problem is how to obtain information to estimate the

[1] *Chemical Age*, October 7, 1966, and January 28, 1967.
[2] See C. Pratten and R. M. Dean, *The Economics of Large-scale Production in British Industry* (University of Cambridge, Dept. of Applied Economics, Occasional Paper No. 3, Cambridge University Press, 1965), pp. 12–22.

extent of economies of scale. Pratten and Dean noted that one could either 'compare the performance of existing plants, or . . . estimate the costs of production for hypothetical plants of varying scale'.[1] They chose the latter method, and we shall also use mainly hypothetical data, since detailed plant comparisons are difficult to obtain and may reflect large changes in techniques. In any case, estimates of hypothetical costs are based substantially on actual costs and are more likely to reflect economies of scale as the economist defines them.

Production cost estimates for plants of different sizes suggest that substantial economies of scale do exist in many sectors of the chemical industry, as a few examples will show. In the manufacture of urea, a component of nitrogeneous fertilizers, a plant of 500 tons per day capacity has a production cost of 9 per cent below that of a plant producing 100 tons per day.[2] The cost of producing ethylene, a basic plastics raw material, is about 20 per cent higher with a plant of 100,000 tons per annum (t.p.a.) capacity than with a 300,000 t.p.a. plant.[3] Nitric acid produced in a plant of 180,000 t.p.a. capacity costs half as much as that produced in a plant of 50,000 t.p.a. capacity.[4] In the production of ammonia it has been estimated that a unit cost of production would fall by 32 per cent as plant size increased by a factor of six,[5] while another estimate based on actual plants was that unit costs would fall by 21 per cent as plant size increased three times.[6] With potential economies of scale of these magnitudes in certain sectors of the industry, two questions arise. First, what are the sources of these economies, and secondly, what effect have they had on the implementation of technological change which might help to explain the effect of technological change on employment?

An important source of scale economies is that large plants use less capital per unit of output. The cost of pipes, vessels and tanks fabricated from steel is determined by their surface area, but surface area does not increase proportionately with the volume which it can

[1] See *The Economics of Large-scale Production in British Industry*, p. 19.
[2] *European Chemical News Supplement*, 'Large Plants: their Design and Economics', October 16, 1964, p. 26.
[3] Ibid., p. 36.
[4] Ibid., p. 13.
[5] United Nations, *Cement/Nitrogenous Fertilizers based on Natural Gas* (UN Studies in Economics of Industry, New York, 1963), p. 38.
[6] *Chemical Age*, January 13, 1967. It was estimated that of this 21 per cent reduction in costs, 10 per cent was due to increased capacity and 11 per cent to improved technology and 'better integration of energy balance within the plant'.

contain. Hence the cost of such containers increases less quickly than their capacity, and the relationship between surface area and volume has given rise to a 'rule' known as the Six-Tenths Factor, which states that where c_1 and c_2 are the capital cost of two plants and v_1 and v_2 their output capacities, then

$$c_1 = c_2 \, (v_1/v_2)^{0.6}.$$

Thus a plant to produce 500,000 t.p.a. would have a capital cost only about 52 per cent higher than one to produce 250,000 t.p.a. Strictly speaking, the Six-Tenths Factor cannot be applied to actual complete plants, since it refers only to certain types of component used in plant construction. Also, it assumes unchanged technology and in a new plant this obviously may not be the case. There may be changes in temperature and pressure which affect the quality or thickness of steel, and much more elaborate control systems are needed as plant size increases. In a very large modern plant, for example, the cost of instrumentation and control could amount to as much as 10 per cent of the total capital cost. For this particular part of capital cost, therefore, increasing plant size could lead to a disproportionate or at least equiproportionate increase.

There are limits to the extent of economies in unit capital savings flowing from larger plants. For one thing, considerable diseconomies result if equipment has to be fabricated on site. Construction of a chemical plant is a matter of assembling and connecting equipment which has been built in a factory and transported to the site, and if any of the individual items become so large that they cannot be transported, the difficulty and cost of putting them together on the site removes much of the cost advantage of their size.[1] There is an additional complication if multiple units are used, for the Six-Tenths rule would not operate if one unit were replaced by two larger ones rather than a single one. However, this does not often happen in practice for most very large modern plants are 'single-stream' i.e. the product flows through the series of sub-processes each of which consists of one stream. If the flow at any stage ceases, the plant as a whole must stop operating. The rate of flow obviously depends on how much product the least efficient stage can handle, but it is not normally considered economic to double-up this stage. The complications of installing and operating part-double part-single-stream plant would mean higher capital and running costs

[1] For this reason, an important limitation on the size of an individual piece of equipment is the capacity of the road or rail system to carry it. Sea transport, of course, is used wherever possible.

than with the largest possible single-stream plant, even though the output capacity of the single-stream plant might be smaller. Notwithstanding these qualifications on the application of the Six-Tenths Factor, empirical estimates suggest that the average value of the scale factor in some sectors of the chemical industry and in oil refining ranges between 0·6 and 0·9, i.e. there can be a considerable saving in unit capital requirements from an increased scale of plant.

Larger plants also tend to have lower labour and land costs per unit of output. This is partly because increasing the size of plant in output terms is often accomplished by increasing the pressures at which the plant operates, the size of pipes, or the rate at which the product flows. This involves no more site area, and the additional instrumentation and control required increases the scope of the process operator rather than the number of operators. Maintenance demands may be greater, but do not usually rise in proportion to output capacity; one estimate is that a doubling of capacity would increase maintenance costs by between 40 and 60 per cent, while most overhead costs do not increase.[1]

In assessing the importance of economies of scale in the implementation of technological change (and vice versa), we may concentrate on one sector where these have been particularly important, namely the plastics sector. Plastics goods were originally substitutes for the more traditional materials such as wood, metal and glass, partly because of their intrinsic advantages of lightness, flexibility, durability and ability to take colour, but also because plastics had some price advantage. As demand increased because of product substitution and new uses for plastics products, technology improved and larger plants were introduced to manufacture plastics raw materials. These lowered unit costs, enabled prices to fall in relative and absolute terms, and further stimulated the demand for plastics products. Hence the consumption per head of plastics materials more than doubled from 1958 to 1967, while their wholesale price fell by about 25 per cent. In this sector, technology is making larger and larger plants possible, large plants exhibit considerable economies of scale and allow lower unit costs, and demand for plastics materials has been growing at an average rate of between 10 and 20 per cent per annum. It is therefore reasonable to expect that the implementation of new techniques, embodied in larger and larger plants, would be very rapid.

[1] Sir Ronald Holroyd, 'Advantages and disadvantages of very large single path chemical plants', Presidential address to Society of Chemical Industry: quoted in *Chemical Age*, July 22, 1967.

In practice, a firm considering the introduction of a new plant has to estimate the rate of increase of demand, and then decide how to phase its investment. There are two related problems here. First, what should the timing of the plant's introduction be: and secondly, can a very large plant operate at a high enough level of capacity? This is not a matter of breaking even, but of showing an adequate rate of return. Should a firm install a very large plant which may stretch technology to its limits, should it construct a smaller version, or should it phase in several smaller plants as the demand warrants? The problem of timing is to phase the capacity installed with the rate of growth of demand. In Britain, a company may spend 9 to 12 months planning the introduction of a large plant, and it would probably take about two years to complete.[1] Its capacity must be sufficient to cope with expected demand over a number of years ahead, so that the new plant would initially be operating well below capacity and therefore at a higher unit cost. This is particularly true since fixed costs represent a high proportion of total costs in large plants with depreciation and interest payments accounting for 20 to 40 per cent of total costs.[2] The question then is whether, instead of one very large plant, it would be more economic to introduce several smaller plants, suitably phased, each of which would work to full capacity within a very short time, rather than one very large plant. However, the economics of large plants are such that this is not in fact the most profitable solution. An example derived from the production of ammonia (in petrochemical-type plants) compared the discounted cash flow rates of return on a 1000 t.p.d. plant and three 333 t.p.d. plants phased to meet demand.[3] If all units were producing at 100 per cent of capacity in Year 1, the rate of return would be 26 per cent for the large plant and 4 per cent for the smaller plants. Over a three-year period with the large plant working at 30 per cent of capacity in Year 1, 70 per cent in Year 2 and 90 per cent in Year 3, it would still show a higher rate of return than the phased smaller units. The same relationship

[1] National Economic Development Office, *Investment by the Process Industries* (NEDO Process Plant Working Party, Report No. 2, London, September 1967), para. 55.

[2] See *Large Plants*, op. cit., and United Nations, op. cit. Depreciation is calculated on a 10-year straight-line basis. It has been suggested (*Chemical Age*, March 27, 1965) that a much higher depreciation charge should be made in the early years of a plant's life, for example that 70 per cent of the capital cost should be paid off in the first four operational years. This would increase the proportion of fixed costs in these early years.

[3] Sir Ronald Holroyd, op. cit., quoted in *Chemical Age*, July 22, 1967.

would hold for other petrochemical plants: any new plant should be as large as the market warrants and as technology allows.

In arriving at this type of investment decision, companies must forecast not only the growth of the market, but their share of that market and the trend of prices. The market for plastics materials is a fiercely competitive one, both domestically and internationally, and in trying to increase the capacity usage of their large plants, companies are all competing for a larger share of a market which, at particular periods, is considerably smaller than the combined capacity of home plants, let alone the import capacity of foreign competitors. This happens because investment plans for large units have tended to bunch, so that several large plants, each anticipating several years' market growth, come into commission at the same time.[1] Thus, because of the introduction of several large new plants for the production of ethylene, United Kingdom capacity in 1970 was estimated to be 1,350,000 t.p.a., compared with 523,000 t.p.a. in 1965 (*Financial Times*, July 29, 1968). One way of counteracting the possible effect of such overcapacity on unit costs, or more generally dealing with periods when demand falls below the expected level, is to 'dump' a proportion of output in any suitable market. To keep capacity utilization high, it is desirable to sell at any price covering variable costs, which might be anything from 15 to 40 per cent below full price, and, since many large chemical plants are being erected all over the world, there have been frequent allegations of chemicals, especially petrochemical products, being dumped in convenient export markets. The British tariff on various types of chemicals has customarily been lower than in other industrialized countries, and the British market has often been a target for dumping,[2] but domestic competition rarely goes as far as dumping or indiscriminate price-cutting. It would be difficult for a company to 'dump' in one sector of the market while maintaining prices in another sector, and though fierce price competition does exist because of the introduction of new techniques which reduce costs, no company is willing to cut prices to an extent that it starts a price war. Competitive price cutting and selling below cost would endanger the whole structure of prices and reduce drastically the profitability of new investment in so far as the actual selling

[1] Section III of this chapter gives further details of investment in the industry.
[2] In 1967 the British tariff against plastics materials was 10 per cent as against 20 to 80 per cent in other industrialized countries, but the Kennedy Round of 1967 proposed that other countries' tariffs on plastics materials should also be reduced to 10 per cent; this should further stimulate international competition.

price fell below the price estimates on which the investment had been undertaken.

So far, we have discussed the factors underlying the implementation of new techniques as if all new investment represented a gross addition to production capacity, but of course this is far from the truth. The reason why technological changes have had employment effects in the chemical industry is that large new plants usually replace smaller plants as well as coping with the growth of demand. If the cost of producing ethylene on a 450,000 t.p.a. plant is half that on a 150,000 t.p.a. plant, it may well be profitable for a company building a large plant to close its smaller plants entirely so as to increase capacity usage of its large plant. There is also the possibility that plants which are only marginally uneconomic could be kept in readiness and recommissioned when the growth of demand warranted but before a still larger plant was economically or technically feasible. Clearly the combination of production and investment decisions will depend on the company's forecasts of market, market share, prices, and cost of capital, but in the market for petrochemical products which has existed through the mid-1960s the trend has been for the most modern techniques to be implemented almost as soon as engineering advances have made larger plants technically feasible.

(iii) *Change in the Industry in General*
The pace of change in the petrochemical sector, especially in plastics materials, has been very rapid. Demand is increasing very rapidly, the best technology of production is improving, the implementation of new technology is not delayed, and the economies of large-scale production allow much lower costs and prices, with consequent effects on demand. How far can one generalize this example to the chemical industry as a whole? As one might expect, the pace of change is slower in most other sectors; though the forces making for change are the same, the stock of technological knowledge is advancing less rapidly, and the economic factors determining the implementation of the change are apparently less favourable. Most important of these are the rate of increase of demand, the extent to which innovations are cost-reducing, the trend in factor prices, and the strength of competition.

In the general chemicals field, the rate of increase of demand is slower than that experienced in the petrochemical sectors. One reason is because there has not been the same product substitution effect consequent on favourable movements of relative prices.

235

The prices of general chemicals have risen only slowly in recent years, as Table 9.3 showed, but plastics materials prices have fallen substantially. Also, of course, the opportunities for product substitution are much greater for plastics than for chemicals used as raw materials in relatively small proportions. The demand for many general chemicals depends on the rate of increase of demand for the products of particular industries or a group of industries in which the chemicals are used. Because demand is increasing more slowly, the effect of scale economies in the production of these chemicals is usually more limited. There are exceptions, though. The output of fertilizers has grown relatively slowly by chemical industry standards, yet to meet overseas competition, British companies have been forced to introduce petrochemical-type processes to manufacture ammonia in plants where considerable economies of scale exist. This defensive investment has meant that the producers of fertilizers are in much the same dilemma as producers of plastics, with the added complication that most of the increased supply from the new large plant will have to be sold abroad.[1] But many general chemicals are produced in relatively small quantities, and while large plants are technically feasible, neither the size of the market nor its growth warrant their introduction. Again, the difficulty and cost of transporting dangerous or corrosive chemicals can limit the introduction of a few large plants to serve a scattered market; a larger number of smaller plants may be more economic, even though their production cost is higher. The introduction of very large sulphuric acid plants, for example, is inhibited by this problem, though there are substantial economies in the introduction of large plants and many small plants have been closed in recent years.

It is difficult to say whether less severe competition in the production of general and industrial chemicals helps to explain the apparently slower rate of technical change. Import figures suggest that there is quite strong international competition in certain branches of inorganic and other chemicals, but imports tend to be very variable, and are often of chemicals which are not made in this country, or for which there is a sudden temporary upsurge in demand which home capacity is inadequate to meet. It is reasonable to expect a higher level of competition in the international and domestic markets for

[1] The market for nitrogenous fertilizers has in the past suffered from overcapacity, and an international price war in 1960–2 which upset the whole structure of prices led to the formation of Nitrex, an export cartel of European producers which was designed to counter the enormous buying power of the state agencies of underdeveloped countries.

standard chemical products which are produced by highly capital-intensive equipment and for which domestic demand is increasing rapidly, though sometimes not rapidly enough to ensure full capacity utilization, than in the market for general chemicals where there are many small plants producing a limited range of chemicals, often tailored to their users' specifications, and serving a localized market. These plants continue simply because their replacement by a new plant embodying a new technique is not economically worth while. Sometimes the operating cost per unit of the old plant as it exists is lower than the total unit cost of the new technique, and often there is again a choice between new techniques. If the market is not expanding quickly enough to allow a replacement plant with a significant net addition to capacity, the alternatives are either to replace the plant by one of almost equivalent size, or to retain the old plant, but to replace and improve parts of it which are particularly inefficient. This latter course is a common type of technological change in small firms or stagnant sectors of the industry. The process may be technologically rather backward, but the replacement of certain parts of it can allow production to continue at a very competitive cost level. This type of small change involving simply an improvement in equipment rather than a complete new plant or even a new process, may have little effect on labour requirements, and any effect it does have is likely to create few adjustment problems.

Any conclusion on the pace of technological change must be impressionistic, but it appears that in some sectors of the industry the stock of knowledge is rapidly increasing and the implementation of this knowledge is also rapid. The major influences causing this are the rate of growth of demand, the cost-saving nature of large-scale investment in new techniques, and competition in the industry. In other sectors, the rate of growth of demand is slower, the installation of large plants is more difficult, and economies of scale are less important in reducing unit costs, though new techniques may be introduced which improve productivity. The difference is the speed with which the process of innovation and implementation occurs. Finally, there are the stagnant or declining sectors where the scope for large-scale investment is small, and where much technological change may take the form of replacing the most obviously inefficient parts of a relatively old process. This type of change may in the circumstances be the most economic, and may allow the old technique to remain competitive with newer techniques installed elsewhere.

III. INVESTMENT IN THE CHEMICAL INDUSTRY

The implementation of technological change comes about through investment, and it remains only to consider the volume of investment in the chemical industry. Table 9.4 shows gross fixed capital

TABLE 9.4

Gross Fixed Capital Formation in the Chemical Industry and Mineral Oil Refining 1957–67

| | Gross fixed capital formation (£m) | | Chemical industry as per cent of all manufacturing industry |
	Chemical Industry	Mineral Oil Refining	
1957	150	42	15·8
1958	157	40	17·0
1959	145	23	16·7
1960	134	15	13·1
1961	167	24	13·5
1962	170	17	14·6
1963	136	27	13·0
1964	174	28	14·3
1965	228	29	16·2
1966	262	57	17·5
1967	215	94	15·0

Source: *National Income and Expenditure*, 1968.

formation in the Chemical & Allied industry (and Mineral Oil Refining) from 1957 to 1967, at current prices. Investment has fluctuated about a rising trend, with peaks in 1958, 1962 and 1966, but investment in chemical plant does not exhibit the 'lumpy' characteristics of investment in the heavy industries, and so the variations in investment expenditure are much less pronounced than in, say, the steel industry. The chemical industry has had investment booms, in 1955–7, in 1960–1, and especially 1965–6, but investment has not fallen very substantially during the off-years. Investment in Mineral Oil Refining has been much more variable and there was very rapid growth in that sector in 1966–7.

Unfortunately, no reliable data exists for investment in the various sectors of the industry, but an annual survey of plant spending is made by *Chemical Age*. This survey covers projects completed during the twelve-month period, those in hand, and those announced during that period for future construction. The investment figures therefore cannot be compared with those of official fixed capital expenditure, since some of the data are estimated by *Chemical Age* or are company estimates of future expenditure. Year-to-year

238

comparisons are imprecise, especially since they involve double-counting in so far as a large project may appear in several successive years, and announced projects may be deferred or cut back. None-theless, the estimates are of interest in showing the sectors where large projects are concentrated, though small investment projects may not be included at all.

Table 9.5 shows the value of projects in hand, completed or planned in the five years 1963 to 1968. Although the petroleum and town gas sectors are not in the chemical industry proper, they have been included in the table, since their investment needs make heavy demands on the chemical processs plant industry. The table shows that total investment more than doubled between 1963 and 1968, with particularly large increases in the petroleum and town gas sectors, and in heavy organics and synthetic fibres. In the 1968 survey these four sectors accounted for 78 per cent of capital expenditure, and, with fertilizers, they indicate the enormous im-portance of petrochemical-type plants in the recent development of the chemical industry. The last column of Table 9.5 shows the average cost per project in 1968. The petrochemical-type plants had a much higher average cost than projects in other sectors of the industry, and the average value of a petroleum project was over £6 million. As might be expected in view of the trend toward larger plants, the average cost per project has been increasing, and the 1968 figure of £2·39 million was some 85 per cent above that for 1963. Only in fertilizers was there a substantial decline. The average value per project in 1968 of £3½ million was little more than half the figure of 1963, a year in which some very large petrochemical-type fertilizer plants were under construction.

Table 9.5 also shows the tendency for investment to be concentra-ted at particular periods as projects bunch together. There was a very large increase in actual or planned expenditure between 1963 and 1965, and a relatively slow rise thereafter. In fact, investment in most of the major sectors in the chemical industry proper declined between 1965 and 1968. The large growth in capital expenditure came in the town gas sector between 1965 and 1967, and in petroleum between 1966 and 1968. This bunching together of investment projects was one of the major problems of the chemical industry in the mid-1960s, since it imposed a considerable strain on the process plant industry's ability to keep a balance between capacity and orders, and led to difficulties in the planning, phasing and completion of investment projects. The investment boom of 1963–5 provides a good illustration of the problem. The process plant industry's

239

Table 9.5

Estimated and Planned Capital Expenditure in the Chemical Industry (projects completed, in hand, and announced): 1963–8

	Value of projects (£ million)						Index 1968 (1963=100)	% value of projects 1968	Average value per project 1968 (£m)
	1963	1964	1965	1966	1967	1968			
1. Heavy organics	76·0	110·3	188·8	188·0	202·5	164·4	216	18·5	2·88
2. Synthetic fibres	29·8	108·3	112·2	103·0	94·5	81·0	272	9·1	4·76
3. Fertilizers	42·9	71·6	104·0	87·8	60·3	32·0	75	3·6	3·56
4. Inorganics	36·9	25·3	55·3	65·3	53·0	45·4	123	5·1	1·19
5. Plastics, resins	24·6	24·0	41·3	40·6	33·9	38·8	158	4·4	1·18
6. Pharmaceuticals	8·9	7·7	13·3	23·5	16·9	12·5	140	1·4	0·89
7. Other	42·8	36·9	46·8	43·5	41·3	43·6	102	4·9	1·11
8. Miscellaneous	19·0	12·5	12·5	15·3	18·1	23·6	124	2·7	0·44
9. Petroleum	66·9	180·2	171·0	162·2	193·8	266·2	398	30·0	6·05
10. Town gas	64·8	67·9	93·0	167·9	215·7	181·2	280	20·4	2·29
	415·5	644·6	838·4	897·3	930·2	888·6	214	100·0	2·39

Notes: 1. Lines 9 and 10 lie outside the chemical industry as we have defined it.
2. Line 7—'other'—includes Dyestuffs, Industrial Gases, Coal Chemicals and Synthetic Rubber.

Source: *Chemical Age*, Chemical Plant Surveys, 1963–8.

normal capacity was sufficient to meet more or less normal investment demands, with additional loads being met by lengthening
order books, and the surge of investment in 1964–5 led to a serious
overloading of the industry. It also resulted in plant contractors,
who fabricated some pieces of machinery but whose main role is to
assemble equipment supplied by sub-contractors, being faced with
delivery dates which were impossible to meet. A special *Chemical
Age* survey in 1965 found that for many types of equipment every
single contract was experiencing delivery delays, usually serious. It
also found that the trend was towards longer delivery dates, and
that the situation was not likely to improve, since most plant
manufacturers were working at or near full capacity, and their
planned expansion was patently inadequate to cope with the chemical
industry's investment intentions up to 1968. The survey indicated
that a major part of the bottleneck lay in the restriction of supply of
certain components, such as towers, vessels, drums, heat exchangers,
etc., because of shortage of capacity in the chemical plant manufacturing industry.

In mid-1966 the Government set up, through the Economic
Development Committees for the chemical and mechanical engineering industries, a committee of inquiry into the shortage of chemical
plant, but in the light of the investment trends of 1960–6, the shortages
of chemical plant are not difficult to explain or perhaps to excuse.
Table 9.4 showed that investment in the chemical industry stagnated
from 1960–3, then rose abruptly in 1964 and 1965. The plant fab-
ricating industry, geared to a level of demand based on the 1960–3
trend, was quite unprepared for the acceleration in demand, and was
unable to expand capacity to meet the demands of the chemical
industry and of the petroleum and town gas sectors. The normal
methods of dealing with investment peaks, lengthening order books
and depending to some extent on imports, were both used during the
1965–6 investment boom,[1] and the Process Plant Working Party was
concerned to find ways in which the delays in plant completions
could be avoided—and, indeed, the normal completion time of two
years speeded up[2]—and imports reduced. One obvious solution was
that the capacity of the plant industry should be increased, and
while it would be unrealistic to expect the plant industry to cope
with unusually high surges of investment, the second report of
the Process Plant Working Party forecast that investment in the

---

[1] National Economic Development Office, *Investment by the Process Industries*
(Second Report by the Process Plant Working Party, London, September 1967).
[2] Ibid., paras. 55–6.

chemical industry in 1967–9 would continue to average over £200 million per annum, while in petroleum refining it would average about £75 million per annum, though on a declining trend between 1967 and 1969.[1] The process plant industry's expansion plans were not geared to this level of orders and the Working Party therefore concluded that the demand for process plant was likely to continue in excess of the supply. However, there was evidence that during 1967 the inflow of orders had fallen away, and that imports were running at a high level. Also the forecasts of capital expenditure contained in an earlier report[2] were regarded by many plant and equipment manufacturers as being much too optimistic, and by 1967 some sectors of the industry were in fact suffering from overcapacity, while others were overstrained.[3] The Third Report of the Process Plant Working Party[4] confirmed their earlier investment forecasts, and estimated that process plant investment would fall by about 10 per cent up to 1970. However an almost simultaneous report by the British Chemical Manufacturers' Association showed some firms with 30 per cent spare capacity,[5] and relatively few new orders coming in. Again high imports were partly to blame for these discrepancies, but the lack of confidence by contractors in the forecasts was disturbing, especially since many of the component manufacturers are firms with other interests such as constructional steelwork and general heavy engineering, and if the demand for these products is more stable and more profitable in the long run, firms might prefer to leave the chemical plant industry rather than keep capacity intact.

Another possible solution is that better investment planning by the chemical industry could avoid the bunching of investment projects and the instability of the load on the plant industry. However, this implies that the investment cycle in the chemical industry is in some sense a result of misinformation or bad planning, and could be rephased without serious difficulty. The evidence suggests that this is not so. Chemical companies tend to make decisions to invest at the same time because their estimates of future demand

[1] *Investment by the Process Industries*, pp. 7–9.
[2] National Economic Development Office, *Future Demands for Process Plant* (Process Plant Working Party, London, January 1967).
[3] The Second Report noted that some companies are hard-hit by the instability of investment, and 'have difficulty in making adequate profits, in keeping a steady labour force and in finding money or confidence for modernization'.
[4] National Economic Development Council, *Investment by the Process and Allied Industries* (Third Report by the Process Plant Working Party, London, May 1968).
[5] Quoted in *Chemical Age*, March 23, 1968.

follow similar lines and show similar trends. Thus it seems that prior to 1963 there was a general tendency to underestimate the growth of the market in the fast-growing sectors, especially plastics. The investment boom of 1964–6 appears to have been caused by the appreciation of the different market situation, as companies began to take a more optimistic view of future prospects. They were optimistic, in as much as they expected the fast-growing sectors to continue growing, but they also appreciated the growing strength of both domestic and foreign competition and the need to produce at lower costs. For these reasons, the large companies had to build large plants, and hence the 1964–6 plants in the fast-expanding sectors cost much more and involved much greater demands on the process plant industry, not only because of the number of plants involved but also because the very large new plants were all stretch-ing engineering technology to its current limits. The large ethylene plants built in 1967–8 could not easily or safely have been constructed five years before, even if demand had warranted plants of that size. Another point on the unpredictable nature of the progress of change can be instanced by the very rapid development of the method of making town gas from light distillate by steam-reforming, which revolutionized gas-making, and turned it into an adjunct of the chemical industry. This was difficult to forecast in 1960, just as it was difficult, in 1965, to forecast what effect the finding of natural gas would have on the demand for process plant in gas-making. It appears, though, that the demand for gas-making plant will fall drastically in 1969–70,[1] which may cause overcapacity in sections of the process plant industry.

During 1968 and 1969, certain developments took place which attempted to avoid the problem of 'feast and famine' in invest-ment demands, and the consequent bad effect on the process plant industry. One was the amendment in 1968 of the Restrictive Prac-tices Act. This enabled the large chemical companies to exchange information on the potential market for certain petrochemicals, and in this way the companies can phase their investment plans—though not obliged to do so—so as to avoid bottlenecks in the process plant industry and the suppliers of components. In addition, ICI constructed a trans-Pennine ethylene pipeline to carry ethylene from Teesside not only to its own plants in the north-west, but also to Shell Chemicals' Carrington site. This allowed ICI to operate its ethylene cracker at higher capacity, and enabled Shell to cover the

[1] *Investment by the Process Industries*, op. cit., para. 32; and *Investment by the Process and Allied Industries*, op. cit.

period when its own additional ethylene demands were not large enough to warrant a large new investment. Finally, a report by the Ministry of Technology in 1969 suggested that both contracting companies and plant manufacturers should be much larger, which would give them the financial and engineering resources to cope more easily with swings in demand. There is, however, no unanimity that this is necessarily the correct solution.

The investment cycle in the chemical industry and its effects on the plant industry are explicable in terms of the difficulty of forecasting both the increase of demand and the development of technological change. The result, in 1965–7, was serious delay in the completion and operation of many large investment projects, and this had two effects. First, the commercial viability of the plants was seriously impaired. If a plant is delayed because of the non-availability of particular components, the construction cost increases and, taking into account the value of lost output, the total cost of installing the new plant will rise sharply and the rate of return will fall. From the longer-term point of view, markets may be irretrievably lost by the delay, for in a highly competitive environment availability of product is all-important. Of course, the lost output may seriously affect the company's own operations, for new and complex plants often produce feedstock or material for other plants on the same site, and completion delays may cause the company to suffer losses through delay or uneconomic running of these subsidiary plants. A further difficulty with new plants which involve technology of unprecedented complexity or large increases in scale is that they tend to suffer teething troubles. The commissioning of the plants and running them up to full output often takes longer than expected, because the engineering and chemical problems cannot always be foreseen. This has also affected the economics of the early life of many large plants, and the liquidity of the operating companies. The second effect of delays in the completion or commissioning of new plant is directly relevant to our present study. If the new plant embodying technological change is designed to replace an older plant, the delays may complicate labour force planning, both in recruitment and training on the new plant and in running-down the labour force on the old. Chapter 10 discusses the labour-force effects of certain types of technological change as they appeared in our case studies, and illustrates the difficulties created by completion delays.

CHAPTER 10

TECHNOLOGICAL CHANGE AND THE LABOUR FORCE

Previous chapters have made it clear that it is difficult to isolate the effects of technological change on the size and structure of the labour force, and also that the case study approach always runs the risk of generalizing unduly from a selection of cases. This chapter attempts to indicate the labour-force effects of technological change in the chemical industry, and it must therefore begin with an explanation of the method adopted and some of the problems involved. Our procedure was to select cases in which an identifiable technological change had had an identifiable effect on the employment, and this gave ten cases in the chemical industry, distributed among six companies.[1] These companies were, broadly speaking, the large companies of the industry, and though some of the factories concerned were small units affected by technological changes in another part of the companies' operation, the immediate employment effects were mostly quite large, and in some instances very large by the standards of the chemical industry. In the process of adjustment of the labour force to technological change the resources and experience were those of the large companies, and this is rather an important point since it means that so far as the chemical industry is concerned, we have little information on the experience of small *independent* companies. From one point of view, this is probably irrelevant, for if a small company introduces a new plant to replace or supplement an old one, it faces the same kind of adjustment problems as a large company would, though possibly on a reduced scale. It will have to transfer workers from the old to the new plant, or hire new workers and somehow reduce the number of existing employees. Small companies may, however, lack the knowledge or resources to accomplish a successful labour-force adjustment especially when the impact on its employment is a consequence of technological

1 Initially, we had more than ten cases, but some were not investigated in detail since they were very similar to some of the ten finally chosen.

245

change in another company, and no detailed information on this emerges from our study.

As well as being concentrated in large companies, the cases were unevenly distributed between sectors of the industry. There were several in Synthetic Resins & Plastics Materials, and in the Chemicals & Dyes sector though this is a large segment of the industry and gaps remain; but there were no cases at all in Pharmaceuticals & Toilet Preparations, Paint & Printing Ink, Explosives, or in the Oils & Fats sector, four sectors which together employed about one-third of the 1967 work force of the chemical industry. The concentration of cases on two or three sectors reflects in part the type of technological change characteristic of the case studies chosen. They were concerned very largely with new ways of making old products, and with changes in the processes of manufacturing basic or intermediate chemicals which require further processing before final sale, or which are used as raw materials in other industries. Another important point is that the technological changes were all different. In the printing and the steel industries it proved possible to some extent to examine the implementation of a particular technological change in more than one company. In the chemical industry, the changes were multifarious, and though there were certain common strands running through them—in particular the introduction of larger plants—the main aspect of the comparison is in the influence of and reactions to similar *situations* rather than similar changes. To avoid undue detail and repetition, the quantitative effects on the labour force are discussed first in Section I, while the remaining Sections illustrate the common threads of labour-force adjustment to technical change in the chemical industry. Section II covers the adjustment mechanisms used in our cases by companies faced with changing labour-force requirements, Section III discusses the planning of labour-force changes, and Section IV deals with certain general issues and the special features of the chemical industry which have facilitated adjustment to change.

I. THE LABOUR-FORCE EFFECTS

Some of the difficulties of identifying the quantitative effects of technical change have already been discussed in earlier chapters, but there are a few problems applying particularly to the chemical industry which affected the amount and quality of information we were able to glean from our cases. First, large chemical sites contain many plants or units, and sections of the labour force are frequently not permanently attached to a particular plant. Part of the maintenance labour force may be under the control of a central maintenance organization with the remainder under the control of the plant on which the employees work. Technological changes on one plant may therefore directly affect only that part of the maintenance labour force which is based on the plant, while the site maintenance force may be little affected, and the effect of technological change in one plant on the central maintenance force may be offset or exaggerated by technological changes on other plants or by other non-technical change. In theory, it should be possible to isolate the precise maintenance demand of a given plant, and changes in the demand over time, but in practice this has proved extremely difficult.

Secondly, changes in company policy may affect manning just as much as technological changes or the requirements of a new plant. For example, there are several ways of organising maintenance of chemical plants. Scheduled maintenance involves a regular servicing of particular parts of plant and machinery, whereas under 'breakdown' maintenance servicing is performed only when necessary. Obviously the size and structure of the maintenance labour force will be different depending on the company's maintenance policy, and a change in policy would affect the labour force independently of technologicalc hange. Interestingly enough, it appears that the results of changing technology do lead quite directly to changes in maintenance policy and to labour-force changes. The labour force may also be affected if the company begins to contract out work which it previously performed with its own employees, and it is becoming common for civil engineering work to be contracted out in this way. The effect of changes in policy on manning and the labour force is of interest in itself, but it may make difficult the identification of distributional changes resulting from technological change.

Thirdly, it is clear that different companies have different 'jobs' for process workers who have the same nominal job title or description, so that men with the same label are liable to be doing a somewhat different set of duties. Chapter 8 showed that ambiguities can exist in the DEP classification of process workers, but a more serious problem is that technological chance could alter the actual bundle of duties without any change in the job title by which the worker is classified. This could also occur with maintenance workers where upgraded semi-skilled men doing a craftsman's job may either be grouped along with time-served men or classed as semi-skilled. Craft labels may conceal great differences in the work a man actually does;

there are many different types of fitter or even of engineering fitter, and a real change in the labour-force structure may occur undetected as men move from one sub-category to another. In this study, we have been limited to the descriptions used by the companies, but a completely accurate measure of changes in labour-force structure would require to examine job profiles in detail so as to obtain a more acceptable measure than present classifications. Whether such a method involving the use of very detailed occupational descriptions would lend itself to acceptable summary or generalization for policy purposes seems rather doubtful.

Within the limits of these difficulties the case studies provided data on the effect of technical change on the number of process workers, on the craft/process worker ratio, and on the distribution of craft workers. Where the effect of technical change on the process labour force can be estimated the reductions were sometimes spectacular. In one case where a coal-based gas-making technique was replaced by an oil-based technique, the number of process workers was reduced from 380 (plus almost 50 supervisors) to 30 (plus 8 supervisors). Again, a comparison of the process manning on two sulphuric acid plants showed a reduction from 47 in the old plant to 13 in the new one which had a production capacity 60 per cent greater than that of the old one. Another technological change involving a multi-product inorganic chemical plant meant a reduction in process manning from about 120 to 60, for a roughly unchanged capacity.[1]

Two other cases presented information of a rather different kind. First, there was the manpower reduction scheme carried out on a petrochemicals site. This case was different from the others in that it only incidentally involved the construction of new plants. After their completion, management undertook an examination of the labour force on the whole site, old and new plants, and was able to reduce process manpower by 22 per cent and maintenance manpower by 23 per cent. In part, the more efficient use of manpower came about because the company realized that the plants could be operated by a smaller number of men, but it was also due to some extent to 'stored-up' technological change. Small improvements in technology had been introduced over a period of years without the theoretical savings of labour which could have been accomplished being made. Instead, the reduction was only accomplished through the scheme, which in a sense compressed several years' labour-saving into one short period. Of course there is no way of telling how much of the reduction was due to revised manning standards and how much to

[1] These figures represent total manning, not manning per shift.

accumulated technological changes, but it seems likely that the revision of manning was more important.

Another case showed how large percentage savings in process manpower could be created by small technological changes accumulating over time, as well as by large changes involving the replacement of an old plant by a new one. This case concerned the manning of petrochemical plants of different sizes and ages. The new plant had about eight times the output capacity of the oldest, yet it employed two fewer process workers per shift and three less per shift than another old plant about one-quarter of its size. This is a most dramatic example of technological change reducing employment, but the same effect can also be seen by looking at the initial and terminal manning of two old plants and the manpower savings over time. At the beginning of its life, one plant had a shift process crew (chargehands plus process operators) of 22. This was reduced to 14, and eventually to 9. The other similarly sized plant began with 15 and was also gradually reduced to 9. These manpower savings on these plants seem to have come from three sources: first, from initial work study carried out when the plant had been run in and was relatively trouble-free; secondly, from the replacement of particular small sections of plant or process which once in operation had proved to require a disproportionate amount of inspection or control; and thirdly, from the development of better methods of doing the necessary work, the kind of improvement associated with continuing work study which changes the way of doing a job without changing the nature of the job itself. Savings such as these follow only from the detailed knowledge of plant operation which comes with time and experience in running the plant. The reductions in manning mentioned above took several years, and in view of the small numbers involved and the existence of other plants which were coming into commission and to which workers could easily be transferred, it is not surprising that few adjustment problems existed.

What would one expect to be the trend of the maintenance labour force in a situation of technical change? The general opinion of those concerned with the case studies was that savings of maintenance or craft manpower were in the short-run unlikely to parallel the very large savings made on the process side. This being so, one can advance the hypothesis that the proportion of craft workers in the hourly-paid labour force would increase, and by and large, the evidence supports this hypothesis. Table 10.1 shows the proportions of craft and non-craft workers in four companies or sites. In three cases out of four, the proportion of craft workers increased, not by a great

TABLE 10.1

The Craft/Non-craft Distribution of the Manual Labour Force
in Selected Companies and Sites (per cent)

	Company X		Site A		Site B		Site C	
	1956	1965	1956	1966	1957	1966	1962	1966
Non-craft workers	78·3	77·7	80·2	74·4	68·3	66·8	58·4	62·8
Craft workers	21·7	22·3	19·8	25·6	31·7	33·2	41·6	37·2

Note: Supervisors are excluded.

Source: Company statistics.

deal in Company X or Site B, but by about 6 percentage points in Site A. The exception was Site C, where the proportion of craft workers fell, but there are several respects in which this case is different from the others. The time period covered was shorter, and the initial craft proportion, 41·6 per cent, was exceptionally high. The 1966 proportion of 37·2 per cent was still much above that recorded in any other case. This exception notwithstanding, there is some evidence to show an increasing proportion of craft workers, and we may now examine the changing distribution of craft workers between trades.

Can we derive any reasonable hypothesis of the way in which particular crafts should expand or decline, relative to craft employment as a whole, when technological change occurs? As plants become more highly instrumented and as very large new plants replace old ones, the proportions of modern crafts—such as electricians and instrument mechanics—might be expected to increase, and the work of the 'black trades', such as blacksmiths, boilersmiths, and platers, to decline. For certain trades, such as plumbers and fitters, the trend is difficult to predict, for the generic trade names could cover significant changes in the type of work done. For example, fitters in an old plant might be doing routine mechanical maintenance, while in another plant they might be doing sophisticated work which entails fuller use of their skills or a different aspect of skill. There is reason to expect a decline in building and civil engineering crafts. The traditional type of chemical plant consisted of a large number of small pieces of equipment housed in a building, whereas new large plants are often not in a building at all, with the mechanical parts of the plant open to the atmosphere and only the control room enclosed. Also, on older plants there was often a good deal of woodwork or brickwork supports where regular maintenance was necessary, and as older plants of this type are replaced by modern units the demand for building crafts will decline. However, it is civil engineering and building work which companies often contract out to outside firms, and an increasing incidence of contracting out would explain part of the decline in the proportion of these crafts.

Table 10.2 shows the distribution of craft workers in three companies or sites, and allows a rather more detailed classification of crafts, though the changes cannot necessarily be attributed entirely

TABLE 10.2

The Craft Distribution in Three Companies

	Company X		Site A		Site B	
	1954	1966	1956	1966	1957	1966
Fitters, Turners	42·5	42·9	45·5	39·2	44·4	41·2
Electricians	12·6	12·2	9·1	14·0	12·4	12·3
Instrument Mechanics	4·8	6·0	8·0	12·1	4·9	8·3
Welders	2·8	3·2	2·7	1·9	3·9	3·5
Boilermakers, Platers	6·3	6·2	NA	NA	7·1	5·3
Blacksmiths	1·1	0·9	2·2	1·9	0·8	0·6
Plumbers, Pipefitters	2·7	3·0	11·8	9·9	6·1	6·1
Other Engineering Crafts	2·4	2·3	3·8	5·1	10·1	14·6
Joiners and Carpenters	7·1	6·4	2·7	2·6	4·7	3·9
Bricklayers	3·3	2·9	2·2	1·1	2·4	1·8
Other Building Crafts	4·6	4·6	2·7	2·6	3·2	2·9
Other	4·4	4·2	9·6	9·9	NA	NA

Notes: 1. NA = not separately identified.
2. In Site A, the total under 'Other' is mainly due to a large number of coppersmiths.
3. In Site B, the 1966 figure for 'Other Engineering Crafts' is open to some doubt.

Source: Case study data.

to technological change. The points of interest here are the relative stability in the proportion of fitters and electricians in Company X and Site B, the large increase in the proportion of instrument mechanics, and the decline in blacksmiths and the building and civil engineering crafts. Other more aggregative data based on sectors of the industry tend to support these conclusions.

The case studies do not allow any precise estimates of the effect of technological change on the labour force, but they do indicate the trends of labour-force changes, and it is clear that in some circumstances technological change can have a quite sizeable impact on the labour force. With the rapid implementation of technological change, it is important that companies should be able to adjust their labour forces both to small improvements in technology and to large changes in techniques, so that the most efficient deployment of the labour force is achieved with the minimum disruption to production and hardship to the workers involved. Our case studies give much more information on short-term adjustments of the labour force and the planning of these adjustments.

II. THE ADJUSTMENT MECHANISMS

Our cases usually concerned the closure of a technically outmoded plant, the disappearance of a certain number of jobs, and the potential redundancy of the individuals filling these jobs: jobs became redundant, and company policy determined whether or not men also became redundant. The companies had to adjust their labour force to the new requirements and in doing so they could make use of a number of devices which would reduce worker redundancy. We shall consider these adjustment mechanisms separately so that the problems and practice of each are clear, and then discuss redundancy which actually occurred as a residual with which none of the methods of adjustment could cope.

(i) *Transfer*

The most common method used was the simple *transfer* of employees to other plants on the same site or in the same local labour market area. Almost every case contained examples of these short-distance transfers, not necessarily to the new replacement plant, but also to existing plants on the same site, and the number of employees involved ranged from a handful to several hundred. Two related reasons explain the frequency and success of direct transfers. First, the tendency for chemical plants in our cases to be bunched together on large sites meant that there was other similar employment nearby; and secondly, in almost all the cases, the closure of a plant took place in a situation of full employment in the local labour market, and the high demand for labour had usually left companies with vacancies on its existing plants. However, where an old plant is running down without a replacement being erected, or where the number of possible transferees exceeds the number of vacancies, some problems could arise, such as the weight to be given to seniority in selecting transferees, or the existence of vacancies in less favourable jobs, a move to which would mean both mobility down the occupational ladder and also a loss of earnings.

Although all the companies attempted to transfer long-service employees into suitable vacant jobs, length of service was never the only criterion for a particular job, and the greatest emphasis was put on a worker's competence. Only in cases of equal merit was seniority decisive, and while seniority did enter into two of the cases as a complicating factor in transfer, it was in rather a special way. In one case of a works closing down, there was an initial demand by the union that any redundancy should be calculated over the whole site and not be confined to the workers on the closing plant. Such a demand, if granted, would obviously involve considerable 'bumping' as long-service workers on the closing plant replaced short-service employees on the rest of the site, with the short-service men being paid off under the last-in-first-out agreement. The company was unwilling to allow redundancy to proceed on this basis, and a compromise was reached whereby long-service men were given precedence in filling any suitable vacancies, which would be internally advertised. This compromise was successful because expansion in the rest of the site and a very high demand for labour in the local area meant that vacancies were continually being created on the rest of the site as employees left. All employees could be offered jobs in other plants though most preferred to take their redundancy payments and leave.

The second case involved a similar situation. The trade unions again insisted that redundancies should be looked at on the basis of the site as a whole, and they also claimed that retention of a particular job on any plant should depend on the worker's length of service *on that plant*, and not his service with the company as a whole. Potentially redundant men from the closing plant would then 'bump' men with shorter service on other plants, and those who would make the last move to the least good vacant job would be those employees with the shortest plant service, even though in some cases they might have long *company* service. The company's view was that redeployment of this kind would require substantial retraining and would lower efficiency in the plants to which workers were transferred. After negotiations, the trade union district officials suggested that management should nominate certain key jobs which they considered to be crucial to the safe and efficient operation of the plants. The occupants of these jobs would be free from redeployment irrespective of their length of plant service. This compromise was accepted and redeployment proceeded on this basis.

When the number of potentially redundant workers exceeds the number of equivalent suitable jobs, workers may suffer a loss of earnings in the jobs to which they are transferred. In three cases, this possibility was covered by *protection of earnings* schemes. In one case, workers moving from a highly-paid job to a lower-paid

253

job were paid at the higher rate for an agreed temporary period, after which they were paid at the rate for the new job; this was a purely *ad hoc* arrangement covering a small group of workers. In the other two cases, definite schemes existed under which employees had their previous earnings maintained (for four weeks, and for one week per year of service respectively), then scaled down (at ten and five shillings respectively) until the new rate was reached. In fact, the number of workers who did lose pay was not large in any of the cases, for most workers either got similar jobs with equivalent earnings, or if redeployed into a lower job were quickly upgraded to jobs with higher earnings.

(ii) *Retraining*

Where a new replacement plant is on the same site as the old one, there may be a problem of manning-up in addition to the run down of the old plant. The new plant may be technologically very different from the old, and the men for a new plant often come from a variety of other plants, including the one to be closed, with most of the men from the old plant being redeployed to suitable vacancies in other plants rather than moving directly to the new plant. Even though transfer usually involves a series of moves of this kind, it is unlikely that any moves can be made without at least some job familiarization and at the most, some quite comprehensive retraining. In fact, the evidence of our case studies suggests that retraining presents few problems for redeployment in the chemical industry. Even if a new plant requires a particular brand of experience which warrants a period of extra training this can be acquired quickly either by a special short-term training scheme or on the job. The process crews for new and technological advanced plants are likely to be small in number and to be selected from the best and most experienced workers on other plants: those who are chosen will presumably already have had some experience of operating advanced plants, so that the period or retraining will be short. The preparation for manning up advanced plants may include a period of theoretical training, practice on training simulators, and detailed study of layout and operation of the plant often uses scale models as well as blueprints and flow charts. At the more practical level, the initial training in plant operation is normally combined with the testing, commissioning and running-in of the new plant. This pre-operational training allows some insight into the likely problems of the new plant as well as giving a very detailed view of the plant in its last stages of construction, and it is most important for the foreman and other supervisors, whose responsibility it is to train other workers, and for senior process operators.

Where men are simply being redeployed into existing plants and not being required to man up a new one, the amount of the retraining necessary for process workers is again likely to be small. In the case studies it was found that those employees who were not directly transferred to similar jobs were fitted into vacancies which required at the most a few weeks' on-the-job familiarization. Again the existence of large sites and the high demand for labour, particularly for craftsmen, made redeployment easier since there were normally vacancies of a broadly similar kind to the redundant jobs in the closing plant. On the evidence of our cases, therefore, the necessity or difficulty of retraining transferred workers is not an important obstacle to redeployment in the chemical industry.

(iii) *Resettlement*

Where there are insufficient vacancies on the site of the old plant or where an isolated old plant is closing down, a policy of *resettlement* may be adopted. Employees may be encouraged to move house to work in a different area where job opportunities do exist, with the company offering financial and other assistance to workers who move. In four of our cases, companies attempted to resettle workers in this way, offering transfer to jobs in a different location with removal expenses paid or substantially assisted. Though the distance of the moves was not great—the maximum was 45 miles, the minimum 12 miles—the attempts were largely unsuccessful. In three of the four cases, almost all of the employees who had originally expressed an interest in moving eventually decided not to move, and even in the other case where more than a hundred hourly-paid workers finally did move, the company estimated that a good many who had initially applied decided not to accept the offer of a transfer.

The general opinion was that there were two main reasons why resettlement offers were not accepted. First, there was the housing problem. In two cases where the move was relatively long-distance, the company did not offer housing, nor was local authority housing available, though the company might have made some financial contribution to the cost of buying a house, while in the one case where substantial numbers moved the company provided assistance in finding housing by arranging with local authorities to make housing available. Secondly, local area ties were considered to be very strong and to prevent even the relatively short-distance moves which some employees had contemplated. Workers simply did not want to leave their home areas, and though in two cases the distance between the old and new plants was so short for there to be a possibility of travel-to-work, employees were unable (or unwilling) to adjust their travel-to-work patterns, and offers of transfer to another plant were refused. The evidence of our cases on the unwillingness of employees to move, especially when a change of residence is involved, is borne out by many other labour-market studies.[1]

The response to resettlement offers was conditioned by the labour market situation in the area. Since many workers could easily find other comparable jobs elsewhere in the vicinity of their homes, the only real incentive to accept a resettlement offer lay in continued employment with the same company. Some employees faced with a resettlement offer considered redundancy, with a high level of compensation, to be a more satisfactory alternative. From the companies' point of view it is interesting to ask why they actually *offered* resettlement, given the cost of transfer. One reason for the resettlement offers was the local labour market conditions in the area to which the transfers were to take place. The economic case for hiring and training 'green' labour was undermined by the low unemployment in the reception areas, and the sites to which resettled workers would have gone were operating with vacancies and high wastage rates. At the time when the cases occurred, specific resettlement offers to suitable workers made some kind of economic sense whereas in different labour market conditions they would not. Nevertheless, in some cases at least, resettlement was admittedly offered as a social obligation and the economic arguments were not examined too closely, though the incentives were not sufficiently attractive for most employees to want to move.

(iv) *Early Retirement*
A particular type of adjustment which can help solve one of the most difficult problems of redundancy is *early retirement*. When a plant closes, the workers who are least likely to be offered or to find new jobs are those who are near the end of their working lives. All the companies studied had pension schemes for their hourly-paid workers, and this enabled them to advance the retiral age for elderly redundant workers. In an unusual form of early retirement applied in a particularly large and complex labour-force reduction, one company simply gave employees close to retiring paid leave until they were able to draw their pensions under the normal terms of

[1] For a summary of some of the evidence, see L. C. Hunter and G. L. Reid, *Urban Worker Mobility* (OECD, Studies in Labour Mobility, No. 5, Paris, 1968).

the company scheme. In another substantial redundancy, employees between the ages of 62 and 65 were offered early retirement, 65 being the normal retiring age. Here the incentive for employees to accept early retirement was not only an immediate pension, but a lump sum based on length of service, average wage and the length of time to retiral, the amount falling the closer the worker was to 65. This lump sum normally varied between £250 and £500. In a third case, employees of 60 or above received an immediate pension, while those between 56 and 59 (and with five years' service or more) received a deferred pension payable at age 60, plus redundancy pay, the level of which again declined the closer the worker was to retiral age. This type of deferred pension for those within a few years of retirement is generally quite common, and the early retirement schemes in our cases were successful within their limited objectives.

In the above examples no question of enforced retirement arose, but obviously it might be in a company's interests to enforce retirement if necessary, so as to reduce the number and cost of eventual redundancies. This type of scheme will operate most effectively where an occupational pension plan exists (though a special form of early retirement scheme could be developed independently of a normal pension plan, or could supplement it), and where the terms are generous, as in the above examples. Early retirement can cope only with a small minority of employees who are close to normal retiring age, but since the problem of these workers is usually the most difficult to solve in a redundancy situation and their re-employment may be impossible, the contribution of such schemes should not be underestimated.

(v) *Redundancy Policies and Payment Schemes*
In view of the fact that our cases all involved quite large numbers of employees, it is not surprising that actual redundancy occurred in most of them, but the success of other adjustment mechanisms meant that in only five cases was redundancy a significant issue. Four of the cases have rather similar backgrounds, and here we shall pick out the common strands of policy, without going into too much detail of schemes. The fifth case contained rather a novel approach to the problem of redundancy, and is considered separately. In the four conventional redundancy situations the main question is whether the companies went beyond the necessary minimum in redundancy provisions on the period of notice, the financial terms offered, and the availability of special terms in the administration of the scheme.

I   257

The period of notice in three of the four cases was considerably in excess of customary practice or of the minimum stipulated by the Contracts of Employment Act. In practice, the length of notice normally varied with service, and the maximum given was twelve weeks which was given in two of the cases to workers with 15 years' service or more. In the fourth case, notice was as laid down in the Contracts of Employment Act. Because of the short advance warning of the precise closure date, the company decided to stick to the shorter period of notice rather than to give long notice based on estimated closure date.

In evaluating the financial conditions of the redundancy policies, it is rather difficult to judge what constitutes generous treatment since the accepted notion of a 'generous' level of redundancy pay has changed considerably over the past few years. The basic formula in all four companies was one week's pay per year of service, pay being inclusive of bonus but exclusive of overtime, but there were variations in each case. In the earliest of the four cases, in 1963–4, the scheme followed the Conservative Government's proposals, giving redundancy pay (up to a maximum of 26 weeks) to all employees of over 26 years of age with over five years' service. In another case which spread over 1964–6, the original proposals envisaged additional payments for service between the ages of 45 and 54, and further additions (declining with age) for service between 55 and 65. The Redundancy Payments Act 1965 caused these additions to be very significantly increased. In this case and another in which the redundancies occurred after October 1966, the company's liability was very much reduced by the repayment from the national Redundancy Fund.

There were other amendments to the basic formula. In two cases, management was concerned that neither the company proposals nor the Redundancy Payments Act gave compensation to the worker with short service. To avoid adverse effects on morale and on the willingness of workers to stay with the company until the works closed, a minimum payment of £50 was given to all workers with one year's service or more. This created anomalies since the £50 payment was not an addition to normal redundancy pay: thus a worker with one year's service and a worker with five years' service on a wage of £10 per week both got £50. The companies decided to accept these anomalies in the interests of giving redundancy pay to short-service workers, and this was particularly important in the case which pre-dated the Redundancy Payments Act, since the minimum service before normal redundancy pay was given was five years. A

258

final type of amendment to the normal redundancy principle occurred in a works running down towards closure through 1964-6. The last two years of plant operation might be untypical and employees might be transferred or have to work under earning conditions which gave a lower level of earnings than their normal rate under normal conditions, so that redundancy pay based on terminal earnings in 1966 would be unrealistically low. To avoid this problem, the company calculated employees' redundancy entitlement up to June 1964, and treated the last two years entirely separately; redundancy pay for 1964-6 was based on earnings at June 1964 or at closure date, whichever were the higher.

The effect of the redundancy on the local labour market was usually mitigated by a form of voluntary or 'early' redundancy, under which certain employees who were able to find other jobs were allowed to leave before the closure of the plant. The mechanics of the system varied between companies. One small departmental redundancy was carried out on a last-in-first-out basis, but the list of those who should be made redundant could be changed if a longer-service worker volunteered for redundancy. If the company could spare the volunteer, he then substituted for the potentially redundant man; about 15 per cent of the labour force were given early release. In another case, employees were entitled to request voluntary redundancy after the closure date had been announced, and employees over 52 could do so at any time. If they had found a job which the company considered to be permanent and suitable and if they could be spared, they were allowed to leave with their redundancy pay and scaled-down pay in lieu of long notice. At October 1966, between 10 and 15 per cent of the hourly-paid workers had taken early redundancy, about 30 per cent of redundancies to that date. The success of voluntary redundancy depends on the ability of workers to find other jobs, and the next section will indicate how a company may be able to plan a closure in such a way that the plant can keep operating with a considerable reduction in employment so that volunteers for redundancy can be spared. The existence of these voluntary or early redundancy schemes emphasizes the difference between redundancy and unemployment, for under these schemes employees drew redundancy pay, often running into several hundreds of pounds, with little or no interruption of employment.

The most significant and successful voluntary redundancy scheme was carried out at a petrochemicals site. Briefly, this manpower reduction scheme involved a very careful study of manning requirements throughout the site, and the surplus by craft and by grade was calculated. Instead of the normal procedure of deciding redundancy by seniority, the company decided to allow employees to leave who had no particular ties with the company. They therefore published the redundancy requirements on the 'first come, first served' principle. For example, if the lists had shown that there was a surplus of ten pipefitters, the first ten pipefitters to volunteer to become redundant would be allowed to go. In order to accelerate wastage, a payment of £50 was made in addition to the normal redundancy scheme, which was already extremely generous, as these details show.

For each year of service below age 30—1 week's pay.
For each year of service between ages 31 and 40—2 weeks' pay.
For each year of service between ages 40 and 59—3 weeks' pay.
The maximum was 50 weeks' pay.

To cater for the low-wage worker and to facilitate the operation of the scheme, all workers with at least two years' service were guaranteed £150.

There are obvious problems about this type of scheme which simply accelerates wastage and gives the company no control over the selection of the individuals who leave. If the initial manning exercise has been conducted efficiently, the correct number of employees will be employed in each craft or grade, but if the workers who decide to leave are the most productive and efficient, the company may find itself with a labour force of lower quality and productivity than it would like, and than it would have been able to keep under conventional redundancy procedures involving selection. In this particular case the exercise worked very successfully, though some workers went whom the company would prefer to have kept. This is not surprising, since the local labour market offered very good re-employment prospects, especially for craft workers, and the financial terms were extremely generous. The average payment was about £250 and the average service of the volunteers was about five years; obviously the average payment of £50 per year of service was considerably greater than most redundancy schemes would have offered. A minor point which cannot be ignored is that the company announced the scheme as an experiment. If it did not succeed in reducing manpower within six months, there would have to be 'another look' at the situation, and this might have involved a conventional redundancy plan, though this was never put forward as a possibility. However within two months the required number of employees had signed forms accepting redundancy

260

and agreeing that they would leave at some date in the near future.

Before the Redundancy Payments Act, voluntary redundancy schemes of this kind could only be carried out successfully where a company was relatively large, profitable and had a record of good labour relations, and where demand conditions were such as to make any further redundancy unlikely. This type of scheme could be of importance in a situation of known overmanning but in the case of a complete closure it would have little relevance. The real lesson of this case is how one company overcame what it thought the real problem of redundancy, namely the selection of the individuals who were to be paid off. By accelerating wastage of those whose ties with the company were loosest, the firm hoped to achieve the manpower savings with 'dignity and humanity', even though the money cost might be higher.

III. THE PLANNING OF THE CHANGES

(i) The Planning of Continued Plant Operation

In the planning of labour-force adjustment to technological change, the most important point is how successfully a company could plan the transfer of workers, their retraining, recruitment, redundancy, etc., in such a way as to minimize dislocations of production and hardship to employees. Although the main problem of the redeployment of labour is usually seen as the movement of labour from one plant to another, another difficulty where an old plant is to be closed is how to keep the plant in operation up to its shut-down after the date of closure has been announced. In most of our cases, the demand for the product was increasing, and an old plant was being replaced by a larger plant which could produce at lower cost. Domestic and international competition in the chemical industry is particularly keen, and to maintain markets it was essential that the old plant operate until the new plant could take over supply, though given the lower unit costs of the new plant it would be uneconomic to keep the old plant open thereafter. This posed some intricate problems of planning and adjusting both production and the labour force, as the new plant would be manning up concurrently with the old plant closure. A number of particular difficulties can be illustrated with examples from our case studies.

One is that the training of supervisors and operatives on a new plant generally begins during the late stages of plant construction and continues through the commissioning and running-in stages. If it is

261

planned to transfer supervisors or workers from existing plants, this will leave gaps in the labour force of the existing plants, and since those who are transferred will tend to be the best and most experienced workers, the existing plants may be denuded of their most efficient employees. In two cases in particular, this kind of transfer caused difficulty, and process workers on the old plants were upgraded to temporary supervisory posts, with temporary workers hired to cover gaps in the hourly-paid labour force. These solutions could have created their own adjustment problems when the old plants finally closed, for by that time there might have been relatively few jobs into which the temporary supervisors could be redeployed as process workers, but once again the existence of large sites enabled these workers to obtain other jobs. Since those who did get temporary promotion to supervisor were likely to have been of above-average ability, they were worth redeploying, and sometimes jobs were kept open for them. This might not be the case with temporary hourly-paid workers who in one case were not redeployed at all, but were simply offered temporary jobs on a short-term basis, without entitlement to continued employment or to redundancy payments. Since workers who took jobs on this basis tended to be somewhat footloose, the problem was more to retain them than to redeploy them.

Secondly, a plant does not need the same labour force in the short-run as it would require if it were to continue production over a long period. There are a number of ways in which the labour force can be reduced, and again these were extensively used in the case studies. One simple method is to close down ancillary departments or particular parts of the works which are not essential to its short-term running; for example, the transport department or parts of maintenance can be phased out by contracting out work in the last few months of the plant's life, and particular cases showed particular examples of non-essential departments being closed. Equally important, the maintenance pattern of the plant can be altered. If a plant is normally closed down for yearly maintenance, it may prove possible to eliminate this during the last eighteen months of its life and so allow maintenance workers to be reduced to the number sufficient for normal wear-and-tear or 'patch-up' work. This too was a feature of our cases, as was the rescheduling of work to do away with 'spare' men. Normally a plant will have some spare capacity in its manning standards to cover for the inevitable and reasonably predictable loss of time caused by sickness, absence and holidays. With closure within a relatively short period, it is possible to reduce the labour force by working increased overtime or scheduling extra

262

shifts.[1] This would therefore allow the labour force to be cut while maintaining full production, albeit at higher unit cost.

A third problem arising in the continued running of an old plant is premature wastage. When it becomes known that a plant is to close, wastage may increase as employees leave to take other more secure jobs. When the local labour market situation is tight and unemployment is low, the workers who do leave are likely to be those who might have difficulty in finding another job quickly; for example, there were several instances of older long-service workers leaving to take up a job offer even though this meant sacrificing their entitlement to redundancy payments. Young workers and those whose re-employment prospects were better were much more likely to stay on until closure and then to look for new jobs. However, in all the cases there was little evidence that the impending closures had much *additional* effect on wastage until quite close to the closure date, even where there was high unemployment in the area. Though all companies gave between one and two years' notice of closure it was only in the last few months of running that excessive wastage became a problem, and then it was quite easily dealt with.

In one case where wastage was already high and expected to rise the company made a 'retention payment', an addition to earnings of some 10 per cent, in an attempt to stabilize the labour force. Although young employees continued to leave, it proved possible to keep the works going with the reduced labour force and uncontrolled wastage did not increase significantly. In other cases, the companies used the terms of redundancy payments as devices to reduce wastage. In two cases, short-service employees who remained until the date of closure received specially generous redundancy terms, and several companies made it clear to employees that workers leaving without company agreement forfeited their right to redundancy pay, which undoubtedly helped to reduce premature wastage among long-service workers. This is not to say that every worker was refused permission to leave and take another job. Long-service workers usually received a long period of notice under the redundancy arrangements, and some companies allowed long-service employees to take other jobs during their period of notice. Nevertheless, companies must have been able to stabilize their labour forces through this control, or at least to keep workers who were essential to the continued running of the plant.

[1] An interesting opinion is that, on a large site, redundancies should ideally take place in the spring, since potentially redundant men can be used as holiday reliefs, and the exercise phased over a much longer period.

(ii) *The Planning of Manpower Changes*

The planning of transfers and redundancy was in most cases highly detailed, and the manpower reduction scheme described in the previous section is a good example of rather a special kind. It involved a thorough study of manpower requirements and a drawing-up of a list of the necessary job-redundancies. The company did not plan the selection for redundancy, since the scheme depended for its success on voluntary redundancy, nor was it really able to plan the financial inducement which would persuade a sufficient number of employees to volunteer, though it was safe to pitch the amount high rather than low. The voluntary redundancy terms only applied to as many jobs as the study team had estimated were surplus: if the company had ten too many plumbers only the first ten to volunteer would be allowed to become redundant, and if an eleventh wished to leave he could do so only by resigning, and would not receive redundancy payments. Thus the planning of job-redundancies by craft and grade ensured that too many workers could not go, unless mistakes had been made in the manning study.

The other more conventional redundancy exercises were also planned in some detail. The redundancy arrangements were widely publicized both inside and outside the firms, and all firms made forecasts of the labour force to show service, pension and redundancy entitlement at the closure date. In one case, as mentioned earlier, redundancy entitlement was protected against a possible drop in earnings during the run-down and closure.[1] As far as re-employment was concerned, there was universal use of the Ministry of Labour's services, including the setting up of special offices to cater for redundant employees. Also, most firms made some attempt to inform other firms in the labour market of their impending labour surplus, though the number of workers placed as a result of these inter-company contacts was very small. In one case of a closure the company offered to organize for its workers retraining courses which might increase their prospects of re-employment. About 140 employees requested retraining, but only about 60 were actually given it, since some requests conflicted with union principles, some types of training could not be offered, and some employees were unsuitable or left before taking training. This retraining covered both craft workers, who usually had refresher courses in various aspects of their crafts, and general workers, whose training courses included

---

[1] It is interesting to compare this with some cases in the steel industry, where the companies protected themselves against a rising redundancy entitlement because of temporary promotions and higher than normal earnings.

vehicle driving and a number of semi-skilled types of job. The organizing of these retraining courses was largely a social obligation on the company's part, since there was little chance of these workers being re-employed by the company.

Several examples of planning of inter-plant transfers on the same site show differences in the level of detailed control which companies exerted over the movement of workers from job to job. The general policy was to retain potentially redundant workers as spares, reliefs or extras, and to fit them in when suitable vacancies occurred, but the methods of fitting in varied. In one case, a system of internal advertisement was used, but this was more useful in allowing transfer of workers before the affected plant had closed down. In another case, the surplus workers were divided into three groups, and assistance was concentrated on Group C workers—difficult to place in employment—who were mainly older workers holding well-paid jobs on the old plant. Seniority and protection of earnings were important issues, and it was in this case that the dispute over seniority occurred and the compromise in the manning of key jobs was agreed (see Section II (i) above). Here, too, the most generous protection of earnings arrangements were made, and retraining was carried out on a more formal basis than elsewhere. Though the necessary training was neither lengthy nor intensive, three separate training courses for all workers of one week each were organized by a foreman appointed as training supervisor for the duration of the transfer exercise. In the end all the Cs were either redeployed or retired, though the process took time. Six months after the closure date, 30 per cent of the Cs were still not in permanent jobs: some were engaged on odd jobs, while others had temporary posts or were acting as reliefs.

Two other examples of planning transfers were more detailed still. One of these was connected with the manpower reduction scheme in which, as we have seen, the most important part of the planning came before redundancy took place, for it was essential that the manning standards of the plant afterwards should be strict enough to obviate the need for further redundancy or recruitment in the short term. While the scheme was operating, the site had to be kept running and this meant redeployment of men to fit in with the new manning standards. However, the company could not pre-plan the precise movement of men from job to job, since it did not know which men would leave nor the rate of leaving, though the pattern became clear as the exercise proceeded. During the transitional period, a pool of spare men was established, and from this pool men could be moved around the site to plug gaps which appeared in the manning of various plants as voluntarily redundant employees left. In the event, the company was able to cover the gaps by using the spare men and by working overtime, even though the rate of leaving was at one period higher than had been allowed for.

The planning of individual transfers was attempted in another case which involved the transfer of certain employees to man up a large new petrochemical plant. The planning of the start-up of a new plant is a highly complex operation, absorbing large resources of manpower, especially supervision, and in this case the plant supervisors had begun familiarizing themselves with the plant about three months before commissioning date. They were then responsible for training the senior process operators and later the other process workers, and supervising plant preparation and testing. The scheduling of these training periods is primarily a technical problem, based on the known characteristics of the plant, the novelties in the design, which may demand longer training periods for some workers, and the likely completion date. To assist in the redeployment of process operators a detailed list of possible moves was drawn up about three or four months before the estimated completion date of the plant. This showed the exact job which management would like each individual to be doing after the redeployment had taken place,[1] and several alternative jobs. This meant that at any one time the company had a network of transfers which would result in the best distribution of the available labour force. Naturally, the list had to be amended and revised; plant completion dates changed, and some employees whose transfer plans had been settled left the company. In fact, turnover was relatively low (less than 10 per cent) especially among long-service men who were likely to be redeployed into senior jobs on the new plants, but only about 50 per cent of employees went to the jobs which, one year earlier, it had been thought they would fill. However the differences were mainly in jobs low in seniority or status: the redeployment plans for the senior operators were largely successful.

(iii) *The Success of Planning*
While it is in general true that the adjustment of the labour force to technological change minimized production and employment dislocation, events rarely followed the course predicted by the companies, and the plans rarely worked in all respects. They normally had to be considerably revised, and one important reason was the

[1] The manpower changes involved more than one new plant, and the closure of more than one old one.

266

delay in the completion and operation of new plants. Most of our cases involved new plants, and not one was completed and operating successfully by its original commissioning date. This was caused not only by delays in construction but by teething troubles during the start-up period when the plants were running-in before full production had been reached. Because the new plants did not come into operation at the expected times the closure of the old plants had in turn to be postponed in order to maintain supply and keep custom and goodwill in the face of competition.

This gave rise to a number of problems connected with the deterioration of the plant. Section III (i) pointed out that the annual maintenance shutdown might be sacrificed in order to allow quicker wastage of the maintenance labour force, but if as in two cases an old plant has to stay in production for nine months or more after the estimated closing date, the lack of any large-scale maintenance or overhaul may cause the plant to need a great deal of patch-up or breakdown maintenance. If long-term spares have not been ordered, it may be difficult even to keep production at a high level. Since completion delays were usually only predictable at relatively short notice, the postponement of plant closures normally occurred after an original closing date had been announced, and the uncertainty which this created—and in more than one case there were several postponements—had obvious and adverse effects on the labour force, particularly in leading to increased wastage but also in lowering morale.

As well as affecting the labour force in this way, the lack of a positive closing date played havoc with plans for labour force adjustment. We need simply mention some of the difficulties caused without going into details of the particular cases.

(a) The increased wastage led to additional importance being placed on retention mechanisms, such as special redundancy terms, or insistence on employees completing their service with the company, though as we have seen, high wastage was not generally as difficult a problem as it might have been.

(b) It proved difficult to give long notice to some workers. In one case, long-service workers were given twelve weeks' notice and had secured other jobs before it was realized that the closure date on which their notice expired was optimistic and would have to be postponed. In some cases, long-service workers got payment in lieu of notice.

(c) There was increased reliance on temporary employees to fill gaps caused by wastage, premature retirement or transfer. In the areas of high employment where most of the cases were situated, this sometimes meant hiring workers at the gate, or turning over the 'hard core' of unemployed on the Ministry of Labour's books.

(d) Some supervisors or operators were transferred to a new plant whose completion date was postponed. This led to the training programme being less well-phased than it should have been, and also to unnecessary use of temporary replacement labour or higher overtime on older plants.[1]

Another reason why adjustment procedures may not go exactly as planned is that employment trends may change. Under normal conditions of operating, a company could assume that labour turnover, and wastage, will increase as local unemployment declines, but it is more difficult to predict what effect labour market conditions will have on wastage in a factory which is to be closed. If unemployment is low, wastage may be high, since other more secure jobs can be found, or it may be low, if employees prefer to wait until they become redundant before taking an easily available job. If unemployment is high, one might expect the closure announcement to be followed by a rise in wastage as employees left to take better jobs as the opportunity arose. Clearly an increase in local unemployment might lead to difficulties in placing workers: one of the cases was conducted as a swift redundancy exercise rather than as a slower run-down because of the currently favourable employment situation. But the rate of wastage may follow an unexpected pattern or vary for reasons other than local unemployment or the company's own actions. In one case in an area of high unemployment, wastage did not rise as the closure approached, and the reason here seems to have been that the inauguration of the Redundancy Payments Act and of earnings-related unemployment benefit made redundancy much less of a hardship. In general the case studies show that the demand for labour in the local labour market, either inside the company or outside, is important in facilitating labour-force adjustments.

How successful was the companies' planning? The evidence is that the adjustment of the labour force to change was carried out with very little dislocation and hardship, and even that efficiency of operation was often sacrificed temporarily in order to ease the

[1] An unexpected increase in the demand for the product may lead to a postponement in the closure of an old plant, and in this case the above adverse effects could occur independently of delays in the completion of a new plant, but this is likely to be more easily forecast and it is within the company's control.

transitional problems which might adversely affect employees. The plans for labour-force adjustment were flexible and capable of revision to meet changing circumstances. Indeed the formulation of a plan and the detailed knowledge which this entailed enabled companies to react much more rapidly and certainly to such complications as plant delays and changing employment conditions. The general consensus among companies was that a plan which had a measure of flexibility was desirable, and given the relatively large number of employees involved in most of our cases, it is no surprise that such plans were invariably present.

However, the relatively trouble-free adjustment to technical change may in part reflect the special features of our cases which would not necessarily be generally applicable, and we must take account of how these features affect our conclusions on the labour-force adjustment to change. A basic point is that the large companies which we studied had advantages over the average company. They were large enough to have specialist personnel departments, they had well-considered personnel policies, they all had redundancy schemes and pension plans, and their general welfare policies were extremely progressive by the normal standards of British industry. This provided an environment in which an attempt to plan the manpower adjustments to technological change came naturally; the background was there and so were the necessary resources and specialist knowledge. The success of this flexible planning depended on a set of circumstances and conditions which again might not exist elsewhere. First, the large companies tended to organize their plants on large sites so that the companies could plan redeployment and transfers on a much wider basis than if the affected plants had been isolated. Transfers were accomplished without movement of residence and without a break of the employees' service with the company, and it was facilitated by the characteristics of the process labour force. Relatively few employees had lengthy and intensive training, and jobs on new plants generally required only a short period of on-the-job training before competence was reached. Also, the tendency in the chemical industry for seniority to be a minor consideration in redeployment, save in questions of redundancy or as a deciding factor where employees are of equal competence, meant that workers' previous experience was put to good use and retraining minimized, and it generally proved possible to slot employees into roughly equivalent earnings categories. In circumstances where a long training period was necessary or where seniority had much more weight, redeployment of this kind would be extremely difficult.

A second factor in the success of adjustment to change was the local labour market situation at the time of the cases. In only one instance was unemployment relatively high (above $2\frac{1}{2}$ per cent) and here the introduction of the Redundancy Payments Act went a long way towards easing adjustment. The tight labour market situations did cause the companies some manning problems as outlined earlier, but they reduced greatly the effect of actual redundancies because of the ease with which the labour force could be reduced by controlled wastage. Other plants of the same company often found that they could obtain labour from no other source but the affected plant, and most employees who were redundant found good jobs.

Thirdly, in the majority of cases the companies were experiencing increases in demand for the product of the affected plant or for other products, so that new plants were being introduced or existing plants expanded, with employment opportunities into which workers could be relocated.

Fourthly, the companies were profitable and could afford the cost of the adjustment mechanisms used. This was especially true in view of the introduction of new large plants and the competitive situation in the industry. It was essential to continue the operation of the old plants so that custom and goodwill could be retained and to secure efficient manning on new plants and existing plants. Any attempt to carry out the adjustment of labour requirements 'on the cheap' would have jeopardized the attainment of these economic objectives, as well as having adverse effects on the morale of the company's other workers and on its 'welfare image'.

These considerations show the rather special position of the chemical industry and of our cases. A small independent company operating in an area of high unemployment would have been incapable of dealing with its labour-force problems as they were tackled here. In fact the chemical industry is becoming more concentrated with the large companies accounting for a higher and higher proportion of the industry's output, but this may have even more serious repercussions on the remaining small companies. If they are affected by rapid technological change in other firms, they may be unable to find the resources to compete effectively, and be driven to cut their labour forces or perhaps go out of business altogether. While it is possible that some of the adjustment mechanisms mentioned above could be successfully used, it is unlikely that they could provide the whole solution, and it seems that this is an area in which public policy might most successfully operate.

IV. SOME GENERAL ISSUES

The increasingly sophisticated technology of chemical plants is having more general effects on the labour force than those so far discussed, and three specific areas are important. First, there is the effect of large single-stream plants on conventional methods of manning and maintenance. These plants are designed to operate continuously at a high level of reliability save for an annual or two-yearly shutdown for overhaul and large-scale maintenance. This has the effect of drastically changing the pattern of maintenance. On old-style plants, each stage of the process consisted of many separate mechanical units, for example, ten or twenty small compressors, and several pumps, heat-exchangers, boilers, etc., at other stages. It was possible and necessary to take a unit out of the process and give it regular maintenance, and a large permanent maintenance crew based on the plant worked their way through the pieces of equipment. But with a large single-stream plant, base-load maintenance is low with the peak demand for maintenance coming during shut-downs. Table 10.3 shows the approximate maintenance labour force on two such

TABLE 10.3

Distribution of Craft Workers in an Old and New Plant

Old Plant		New Plant			
		Base-Load		Peak-Load	
Fitters	80	Fitters	16	Fitters	20
Platers, Plumbers,		Electricians	1	Platers	15
Welders	20			Plumbers	20
Electricians	4			Welders	5
Bricklayers	6			Painters	10
				Other civil	8
				Other	40
Total (about)	110	Total	17	Total (about)	120

Notes: 1. These totals include craftsmen's mates, but not supervisors.
2. The 'Other' total in Column 3 consists mainly of laggers and riggers.

Source: Case study data.

plants in one of our cases. The old plant (a coal-based process) had a very high continuous maintenance load requiring over 100 craftsmen, most of them fitters. The new plant (a petrochemical type) required only 16 fitters (four per shift) and an electrician to carry out regular day-to-day maintenance, plus possibly the services of an instrument

271

mechanic when required. The regular maintenance force was therefore reduced to about one-fifth its previous size, but there was a very heavy peak demand at particular times.

The planning of maintenance during the shut-down is very complex, and involves a build-up and dovetailing of the efforts of particular crafts (e.g. plumbers, welders and painters) to minimize the time and cost of the maintenance programme. The demand for particular skills is therefore irregular during the maintenance shut-down, and more important, the demand for maintenance craftsmen on that plant is irregular throughout the year. If a particular plant needs most maintenance workers only for one or two weeks in the year, how should the overall maintenance labour force be organized? On a large site with a number of technologically advanced plants it would appear that each plant should have only a small number of workers attached to it, with the great majority of maintenance workers under the control of a central department supplying workers to plants as they need them for either break-down or shut-down maintenance. If most maintenance workers are presently organized on a plant basis, there are obviously potential difficulties of planning and adjustment to be overcome before a centralized system could operate efficiently. On the management side, plant engineers previously used to controlling their own plant maintenance force would be reluctant to yield their establishment to a central department, and central control might be less congenial to the workers than a plant-based system. It might lead to lack of security and loss of identification with and loyalty to the plant; employees might need to be more versatile and of above average ability if they were to be continually moving around the site doing different tasks; and if the employees did not know all the jobs as well as their present ones, they might fear a loss of bonus or a drop in earnings. These difficulties would only be short-term, but unless planned for they could endanger the economic advantages of centralized maintenance.

A second consequence of the introduction of technologically advanced large plants is their effect on methods of manning. Conditions on these plants are such that conventional job allocations are unnecessarily restrictive and inefficient, and some companies are moving towards a different pattern of manning. This takes as its starting point the fact that the plants are so highly instrumented and so reliable that most of the regular process and maintenance jobs are simple and routine, and goes on to suggest that process workers could carry out minor unskilled or semi-skilled 'craft jobs' while craftsmen could perform some of the lesser duties of 'process jobs'.

272

This idea of production teams acknowledges that the full training of both types of worker would still be necessary on some occasions for efficient and safe plant operation. The intention is not to create a new 'chemical worker' category but rather to redefine the margins between craft and process worker jobs. The concept of the production team is part of a general attempt to improve manpower usage in the chemical industry, and several of the important productivity agreements have attempted to move in this direction. Chapter 11 discusses the importance of manning standards.

Finally, the experience of our cases suggests that technological change is unlikely to create substantial redundancy in the chemical industry, since the adjustment devices discussed above should be able to cope with most labour-force effects. Possibly more important in their effects on the labour force will be the small manpower savings through the marginal labour-force reductions which go on continuously and through bargaining over manning and conditions of employment. The main problem of technological change which the chemical industry will face is not how to manage the release of workers whose abilities are no longer relevant to the industry's requirements, but rather how to obtain workers of the necessary skill and training to fill the jobs created by advanced technology. Some information of the future labour-force trends in the industry came from a report on the manpower requirements of the industry from 1967 to 1973. The report made some estimates, based on a sample of companies, of changes in manpower requirements, and these are very useful in providing an indication of the likely trends in labour demand though they do not pretend to be a statistically accurate forecast. By far the biggest increases were forecast for trained process workers and for technicians, while the largest decline was of ancillary workers, e.g. warehousemen, transport drivers, and other unskilled and semi-skilled workers. The trend of demand for electrical and instrument craftsmen was also strongly upward.

Both the case-study companies and those in the other study were aware that recruitment and training procedures as set up in the past would need some revision. For operators on technologically advanced plants it will be necessary to have more formal training procedures such as the Qualified Chemical Operator's course, and some companies have restructured their training procedures to give a more sophisticated and theoretical foundation to the practical knowledge gained on the job. One problem of this upgrading of training is the educational qualifications of the process worker recruits. A working party from the Chemical EDC which visited chemical plants in the north and east United States and Canada made the following comment:

'Another thing which struck us as being significantly different from British chemical industry practice was the fact that all the companies we visited in North America aimed at a seemingly high minimum educational standard . . . in effect, apart perhaps from such jobs as janitor and cleaner, the chemical industry in this part of America was almost closed to new entrants without this minimum standard.'[1]

If a higher standard is to be required of initial entrants to the industry—and this was put forward as a problem for the future rather than the present even in the advanced sectors of the industry—there are two possible sources of recruitment. One is more highly educated school-leavers who would be able to take full advantage of more sophisticated training procedures. The other is recruitment of scientifically or technically qualified personnel to fill the top process jobs: i.e. the technical white-collar grade might be extended downwards to include the responsibilities of the present senior process operators.

Some people in the chemical industry claim that this latter trend is much more likely. The present practice in the industry is for supervisors to be promoted from the ranks of the hourly-paid workers, and some of our discussions suggested that on advanced equipment the days of this procedure are numbered. Process supervisory posts will be more adequately filled by graduates or technically qualified workers, while the increasing complexity of plants will also demand a higher standard of technical knowledge from craft and maintenance supervisors. Obviously in so far as there was a change in the sources of recruitment to supervisory or top process jobs with technicians replacing hourly-paid workers, this would involve more than a change in nomenclature. There could develop a serious problem of incentive and promotion structure for existing blue-collar workers, for whom the foreman's job is presently the highest grade on the normal ladder of promotion. Again this would only be a longer-term problem affecting parts of the industry, and it might be avoided by a change in the quality of entrants and the method of training.

The important point, though, is that those on both sides of industry are aware of the potentially troublesome areas in the rapidly

[1] NEDO, *Manpower in the Chemical Industry*; Report by a Working Party of the EDC for the Chemical Industry (HMSO: London, 1967), pp. 26–7.

advancing sectors of the industry. Our study of the adjustment procedures and the effect of technological change in the industry suggests that any problems can be solved through a controlled process of negotiation without abrupt and costly dislocation.

PART FIVE
SELECTED ISSUES

CHAPTER 11

MANNING STANDARDS AND WAGES

I. INTRODUCTION

One of the benefits of technological change is that it generally makes possible a reduction in labour requirements per unit of output, thus increasing the efficiency with which a scarce factor of production is used. But the earlier chapters, and other studies, have suggested that the manning standards applied to new equipment do not always reflect the full labour saving potential of the plant. It is, therefore, worth examining more closely how firms can establish and maintain manning standards which are close to the 'optimum' level. Sometimes management will be able to determine the matter unilaterally, particularly where the trade unions are weak or non-existent; though as we shall see, even this does not preclude inefficient working practices from developing. More usually, however, manning arrangements are the result of negotiations in which the company gives certain wage or other concessions in return for the union's acceptance of particular working methods.

This idea of systematic negotiations over matters which many firms have traditionally regarded as managerial prerogatives is relatively new in the British context, though it has gained wide acceptance in the last few years. It stems basically from the view that the inefficient use of labour is a management responsibility, and that they must take the initiative to secure its removal. Furthermore, it is argued, to secure the major changes in established practices which will often be required for greater efficiency, it will be necessary to have the agreement of the workers involved. In order to secure this agreement, the firm will, in most situations, have to offer some tangible concessions in return. The concessions may take various forms, such as greater job security or extended leisure time, but will usually include some increase in wages as well. The questions of manning standards and of wages have thus become closely related, and it is therefore appropriate to consider them together in the present chapter. In some cases negotiations on manning and wages will follow conven-

279

tional collective bargaining lines. But elsewhere revisions to inefficient manning standards have been achieved through what has come to be known as productivity bargaining. While by no means confined to situations where new techniques have been introduced, the concept has in a number of cases studied proved useful in facilitating the full adjustment of the labour force to new working arrangements, often some time after new plant has been installed.

But this link between wages and manning standards is only one aspect of the relationship between technological change and wages. For example, there is the effect of technological change on the average wage level in a plant, and the problem of income maintenance for those transferred to less well paid jobs. Then there is the position of wage incentive schemes, which become less satisfactory means of payment where new techniques remove from the worker the power to control the quality or quantity of output. Finally, technological change may stimulate moves towards new methods of payment giving greater income stability; for example by paying on a salaried rather than weekly basis. All of these constitute aspects of the adjustment process, and must be considered along with the productivity agreements which have a more direct influence on manning levels.

The first problem of this chapter concerns the apparently simple idea of an 'optimum' level of manning for a particular plant. Yet this depends on assumptions or expectations about such things as product quality, maintenance arrangements or the amount of ancillary labour required, and in consequence is a rather elusive concept. But even if it can be established, several pressures are in operation as a new plant is installed which tend to produce over-manning. In part they reflect management policies, and in part the activities of the trade unions. These two questions, the idea of an optimum manning standard and the ways in which overmanning can arise, are the subjects of Section II. In Section III, we look at some of the broader wage effects of technological change. Finally, in Section IV we examine the role of productivity agreements in determining manning and wages, particularly where technological change has led to certain major changes in labour-force structure.

II. MANNING STANDARDS

(i) *Problems of Determining Optimum Manning Standards*

Most of the problems over manning standards for new equipment arise from the degree of discretion which is available in their determination. A particular manning standard has both quantitative and qualitative elements. In both respects there is often considerable discretion in the amount of labour to be used, in the sense that a firm can decide on, or agree to, one of a number of possible manning levels for a particular operation. Given certain assumptions and requirements, such as product quality or the level of maintenance, it can calculate the total number of man-hours required to operate a plant for a given period; this can then be converted into the number of men it will be necessary to employ on the basis of the normal working week and the expected number of days upon which men will be absent due to sickness, holidays and so on.

But although this may establish the optimum manning for one company's plant, it may be an inappropriate standard for another company to use, in so far as its assumptions and requirements differ. For example, the second company may want to produce a product of higher quality and may need to employ more skilled operators or more inspectors in order to achieve this. Another company may place considerable importance on being able to put right any break-down as soon as it occurs, and will therefore carry a very high maintenance staff to meet almost all eventualities. Again the difference between a policy of scheduled maintenance and one involving 'breakdown' maintenance will affect maintenance manning. Such differences in policy, as well as in the proportion of maintenance, building or civil engineering work done by the company instead of by outside contractors, can obviously produce divergences in the manning of apparently similar plants.

The scope for discretion in setting manning standards also depends on the type of machinery or operation being considered. For some machines the manning, at least in quantitative terms, is determined by physical conditions; for example, only one worker can operate a typesetting keyboard at any time. But on larger and more complicated pieces of equipment, the manning standard is less closely specified by the machine; for example, a steel rolling mill will employ a certain number of men, but it is much more difficult to be sure that this is the right level of manning than in the keyboard example. Furthermore, many operations of an ancillary nature to the main process, such as labouring or storekeeping, have an output which it is very difficult to measure. It follows that it is difficult to establish the correct level of manning on these duties in relation to the level of output expected from the main plant, even though the manning of machines may be predictable within relatively narrow limits.

While in theory problems and uncertainties of this sort would tend to produce manning levels both above and below some 'optimum'

281

level, in practice one expects the result to be overmanning rather than undermanning. If a plant is undermanned this will be quickly brought to the firm's notice either by the trade unions arguing that their members are having to work too hard, or by breakdowns and delays in the operation of the plant. As a result, the manning will probably be revised upwards. For example, in the new continuous-casting plant at Appleby–Frodingham, it was found that the original manning levels prevented adequate communication between members of the operating team, who were working at quite widely separated locations on the plant. To improve efficiency, it was necessary to supplement the team. But if a mistake is made the other way and a plant is overmanned, the firm may not be aware of this for some time. Particularly if the plant is of a novel design, there may be no comparable equipment which the firm can use as a standard against which to measure its own manning, and if labour productivity has been improved in comparison with an older plant, satisfaction with this may disguise the fact that it could have been improved still further. Even if the firm later becomes aware of further potential reductions, it may not be able to reduce manning due to trade union resistance.

This leads us to the final difficulty which firms face in establishing an 'optimum' level of manning, namely that the employees or their trade union often have a different view from that of management of what constitutes a desirable manning standard. The level of productivity implicit in the employer's calculations may only be obtained by the expenditure of increased individual effort with fewer breaks or perhaps less freedom to vary the pace of work. These more rigorous conditions of work may be unacceptable to the workers and their trade unions. As other writers have pointed out,[1] to confuse what workers *can* do, on the basis of some form of work study, with what they *should* do, is to confuse measurement with personal judgement. The optimum or most efficient level of manning, and therefore the degree of overmanning in any situation, can be the subject of calculations based on quite different assumptions about such matters as a reasonable pace of work. One would naturally expect, therefore, that the calculations will produce different answers, from which discussions or negotiations can then proceed.

But while it may be difficult to establish precisely what is an optimum manning level for a particular piece of equipment, this does

[1] Ivar Berg and James Kuhn 'The Assumptions of Featherbedding', *Labour Law Journal*, Vol. 13, No. 4 (April 1962); reprinted in '*Featherbedding and Technological Change*', edited by Paul A. Weinstein (D. C. Heath and Co., Boston, 1965).

not preclude observation of cases where overmanning has obviously developed. This can arise in various ways and these are worth examining in some detail in the following section.

(ii) *Reasons for Overmanning*

A firm which has installed a new plant, particularly if it is large and complex, often finds that it is necessary to employ additional men for a temporary period during the running-in period, and there is a danger that at least some of this temporary element may become part of the normal team. For instance, it will be necessary to train supervisors and operators to work the new equipment, and it may be difficult to predict in advance how long this will take. Either the plant is new and its skill requirements not yet properly established, or it may not be known how long it will take to adapt the skills of employees to the new requirements. In the meantime, it will be necessary to have more men on the plant than its intended manning.

Another reason for employing additional men during running-in is to ensure that all the separate parts of the plant are operating correctly and to make any necessary adjustments. There may also be technical difficulties due to faults in design or erection which make the running-in period longer than expected.[1] The fundamental problem seems to lie in the very process of scaling-up a new idea to a commercial scale. Although the theoretical basis of the design of a new plant or piece of equipment appears sound, and trial plants have worked satisfactorily, this in itself does not guarantee that the full-scale plant will work satisfactorily. It seems that quite unforeseen problems can arise once a commercial-scale plant is built, which can only be solved when it is in operation. An additional difficulty lies in the closely integrated nature of many modern processes, which means that a fault in one small piece of equipment can affect the overall production of a plant.

Provided the additional manning required to overcome problems of this sort is reduced as training and running-in problems are completed, no long-term inefficiency results. But there was some evidence in our cases of this subsequent reduction not taking place, and of at least part of the temporary manning becoming built in to the normal complement. Without very close supervision, particularly where the running-in period is longer than expected, working patterns and routines can easily become established which are difficult to

[1] For an estimate of the relative importance of the causes of running-in difficulties in the chemical industry, see *Chemical Age*, July 22, 1967, p. 11.

remove without a further major work study exercise. A rather different, but related, problem has arisen in both printing and steel, where cases were observed in which a new plant was established and manned to the expected level, but product demand failed to build up as expected. As operators finished training, they therefore started working at a level well below their expected capacity. The firms were seriously concerned lest these low output levels should become established as normal, so that when demand did build up they would be unable to meet it without additional labour.

As well as the introduction of major changes in technology, firms are continually introducing small, detailed improvements to plant and small changes in working method. Individually, these may have almost indiscernible effects on the amount of labour required and no adjustment to the labour force is made. But if a succession of changes is made over several years without adjustment of the labour force, this could clearly result in a plant carrying many more people than necessary for its efficient operation. For example, a new plant may be installed, incorporating improved instrumentation and control mechanisms which reduce the process operator's work-load. He may, for example, now be able to control the plant from a central point instead of having to go and turn a valve, saving perhaps 20 per cent of his time. Clearly no adjustment can be made in this one case, but when a succession of small changes of this sort have been made it may become possible to reduce the number of men employed. Alternatively, one can say that the original job organization has been maintained despite the change in content; by changing the job structure it would be possible for the firm to continue to use the man's time fully. But unless a continuous process of job analysis is in operation, the firm may be unaware of the excess which has gradually arisen. A firm will often not make the changes which are possible, as individually they may not seem worth bothering about. If they are allowed to accumulate over time, however, substantial overmanning can build up.

A third source of overmanning is that which results from trade union policies in support of the employment prospects of their members. One of the objectives of a trade union is to maintain or increase the employment opportunities of its members though, of course, the weight given to this in relation to other objectives will differ between unions. The greatest weight is probably given to policies of this type by unions whose members' jobs are being obviously threatened by economic or technological changes, especially where their earnings in other occupations or industries will be less

284

than their present level. For example, the introduction of LD and Electric Arc furnaces into the steel industry has reduced that industry's demand for bricklayers who were previously employed to reline open-hearth furnaces and this has been the cause of several demarcation disputes between the AUBTW and the Confederation.[1]

A union trying to defend its members' jobs against technological change may adopt several courses of action. The first is to delay as long as possible the negotiation of an agreement governing the introduction of the new equipment. An example of this was the delay in reaching an agreement on the installation of mechanical publishing equipment in newspapers, which was at least partly due to the attitude of the union concerned. The AUBTW, as already noted, has pursued a similar policy in steel to delay the rate at which its members are replaced by Confederation workers on the work of lining converters. Although each case they have contested has gone against them, they have continued to claim the work in new installations. Clearly this can only delay the arrival of the adjustment problems (and may make them more sudden and difficult in the end), but it may serve the purpose of allowing the union to start taking steps to meet the new situation.

A second approach is to negotiate manning standards which are higher than the optimum level, in either quantitative or qualitative terms. Agreements requiring the employment of more men than are strictly necessary are fairly common, but have been particularly apparent in parts of the printing industry. But protection of jobs can also be based on qualitative factors, for example, when a union defends its members' position on the basis of a demarcation agreement. A great many jobs in industry have in the course of time, by formal or informal agreement, become the preserve of a particular type of worker. A union may have succeeded in excluding either workers of other skills or workers without skills from doing certain types of work and this allocation may have been tacitly accepted by management. But some types of technological change may so drastically change the pattern of duties as to cast doubt on the traditional allocation of jobs between craftsmen, process workers, and white-collar workers. If a trade union successfully maintains the existing job allocation and, for example, retains for its skilled members a job from which most of the skill has been removed, then clearly an element of overmanning is involved in the qualitative sense. Furthermore, an artificial distinction or lack of flexibility between two groups of workers may lead to inefficient operation if at certain times

[1] A discussion of this problem is contained in Chapter 7.

one group is under-employed but cannot be reallocated to other work.

A third policy which a union protecting its members' jobs may pursue is the negotiation of agreements which prevent a firm from declaring people redundant as a result of technological change being introduced. In contrast with agreements on manning which may preserve jobs indefinitely, 'no-redundancy' agreements only provide security for those currently holding the jobs, rather than for a wider section of the union's present or future membership. This means, of course, that the firm only gradually secures the full labour savings from the new technique. The seriousness of this depends on the level of labour turnover. If the market for a particular type of labour is strong, then surplus labour may leave fairly quickly, particularly if the extra employees enable the firm to reduce the level of overtime working.[1] But one would generally expect the level of turnover to fall in this situation simply because turnover rates are higher amongst short service employees: clearly if no new labour is being taken on the average length of service will increase and the turnover rate tend to fall. If this happens then the wasteful use of labour may continue for a longer period than the firm originally anticipated.

Policies of the sort outlined above will reduce the labour-force effects of new techniques in two ways. Firms which have introduced a technological change will be unable to reduce their labour force by as much as they would have wished. But in addition, by reducing the rate of return on an investment in a particular type of new equipment, employment protection policies will delay, even if only marginally, its rate of introduction into the industry. Some firms will continue to install the new techniques, but in firms where the decision is otherwise finely balanced the higher operating costs as a result of trade union policies will prevent its adoption.

More generally, the effect of trade union policies on the rate of introduction will depend on how important savings in labour costs are in justifying this particular new technique. If this is the main or only reason, then its introduction may be severely delayed. But if the benefits are in terms of better quality or greater output, the additional labour costs as a result of these policies may not affect the rate of introduction very much. Again the extent to which a trade union is able to influence a firm's manning decisions and to protect its members' interest will obviously depend on the nature of the factor and product markets facing the firms. Policies of the sort we

[1] On the other hand, 'no-redundancy' agreements are less likely to be demanded by the trade unions if the market for their members is strong.

have been discussing involve the firm in the employment of more labour at current wage levels than it would if it had unilateral control over the matter. Consequently, if a trade union is to enforce a policy of this sort it must be able to exercise sufficient control over firms' labour supplies to ensure that no serious by-passing of the agreement is possible. If firms could recruit non-union labour on a significant scale then the ability of the union to protect its members' jobs in this way is relatively limited. The success of the printing unions in this respect is, in fact, based on their strong control over the entry of labour to the industry and to some extent of its movement within the industry. They can, therefore, ensure that all but an insignificant minority of firms employ union members according to manning and other agreements which have been reached.

In the product market, two conditions which seem necessary to the successful defence of jobs are either that the industries or firms concerned are relatively isolated from competitive pressures or that labour costs are small in proportion to total costs. Union policies of this sort will obviously raise the operating costs of the firms in the industry. If some companies, or those in particular parts of the country, are subject to more severe demands on manning levels, their relative competitive position may deteriorate. One of the reasons put forward to explain the movement of printing companies out of the London area in recent years[1] has been that the unions had secured higher manning scales there than in the provinces, although clearly the other additional cost of operating in London must have contributed to these moves. A union will probably therefore be better able to secure its position by trying to get similar manning arrangements adopted on a national scale. But if, as a result, the industry is unable to compete with foreign producers or other domestic industries providing substitute products then, at least in the long run, these policies will not be successful. The newspaper industry, for example, is in a very protected position and many firms have remained profitable despite the overmanning which has developed.

But it is also true that even in an industry facing severe competition, a strong union can successfully pursue job-protection policies if labour constitutes a small proportion of total costs. This is particularly true in the case of a small group of workers. Thus a small, well organized craft-union in a key position in the production process may be able to secure adequate employment for its members very easily, simply because they could disrupt production at a cost to the firm far in excess of the cost of concession, especially if all firms in the

[1] *The Cameron Report*, op. cit., para. 104.

industry are affected. In this situation a firm may be more inclined to give way on manning than to suffer the more serious loss of a stoppage of work. This is strengthened by the observation that the main reason for the introduction of many technological changes is not that of saving labour. In these circumstances, firms may well place less importance on securing rigorous manning standards than on getting the plant into operation.

We can see, therefore, that when a firm installs new equipment there are many ways and circumstances in which overmanning may be built into the plant from the beginning, or may gradually arise during its subsequent operation. If the firm becomes aware that this has happened and decides to raise the level of efficiency, the question then is how the manning standards can be changed. As we argued in the Introduction, it will often be possible to make a real impact on the situation only by some form of productivity negotiations with the trade unions, including wage concessions in return for a revised manning arrangement. Consequently technological change may indirectly lead to the negotiation of a productivity agreement, perhaps some considerable time after the new plant or equipment was introduced. But technological change can also have certain more direct effects on wages, and we shall consider those in the following section, before going on to the role of productivity agreements.

III. WAGES AND WAGE PAYMENT SYSTEMS

(i) *The Wage Level*

The first of these wider questions must, of course, be the effect of technological change on the general wage level in a plant. One is immediately faced by definitional, conceptual and practical problems of some magnitude, even though there are good *a priori* reasons for supposing that technological change will in general mean an increase in wages. For example, there is the problem of which workers are to be included in the comparison if there has been movement into and out of the plant during the period of the change. Some workers will have moved, possibly being downgraded, to other parts of the establishment, and a comparison of the overall effect on earnings will have to take them into account. Then there is the question of how we take into account changes in the structure of pay—for example there may be a change in the stability of earnings and, given the importance of this to workers, some allowance will again be needed. Over what time period should the comparison be made? In the short run a displaced worker's previous earnings may have been guaranteed by the firm, but the change nonetheless may have damaged his long term earnings prospects. Finally, there will have been changes in the job content and the skill structure of the labour force on the plant, which will further complicate a true comparison of earnings. From this brief indication of the problems it will be clear that it is difficult to sort out the effect of technological change from all the other reasons why wages might change, and indeed empirical information on the effect of technological change on wages is not very satisfactory.

In our three industries, there were a number of cases where some wage effects could be observed, and the main conclusion is that where there was an old plant with which some comparison could be made, wages on the new one were generally at an enhanced level. In the steel industry, for example, the top earnings set for the new plant tended to be rather above the previous rates, and this seems to have been due partly to higher productivity and partly to the necessity of attracting the best workers to man the new plant. Wage increases have also occurred in cases where small technical developments led to a reduction in costs which was then explicitly shared out between workers and management. Thus when stacking machines were introduced in one printing firm, the division of the resulting cost reduction meant an increase in wage rates of about 4 per cent. In two other printing cases, this gain-sharing approach led to varying increases in the wages of the men whose work was affected. Wage increases of this type are strongly influenced by productivity increases since the increase in productivity allows precise estimates to be made of the reduction in costs. This approaches the productivity bargaining concept which we shall discuss later: We may simply note here that at this level productivity can have a powerful effect on wages. There is some other evidence on the effect of particular technological changes on wages. An OECD international study[1] covered seven cases from Britain, all from different industries, and the level of wages after the change was generally higher. For example, in the chemical company case, there was said to be 'clear evidence that salaries and wages have increased as a result of the development and modernization scheme'.[2]

On the basis of this limited amount of evidence, it does seem that the average wage on a technologically advanced plant is likely to exceed that on a less advanced one, and that employees moving to such plants will receive higher earnings; it may be useful to summarize the reasons why those higher earnings occur. These cannot all be

[1] Solomon Barkin (ed.), 'Technical Change and Manpower Planning', *Co-ordination at the Enterprise Level* (OECD, Paris, 1967).
[2] Ibid., p. 240.

fully demonstrated from our cases, but they do reflect the impressions which companies gave for the higher earnings obtainable after a technological change. First, the productivity argument is significant: since a new piece of equipment almost always reduces labour cost per unit at ruling wage rates, companies are willing to allow increases in wages on the new process. Secondly, the results of a job evaluation exercise may result in higher earnings levels. In both steel and chemicals, for example, considerably more weight is normally given to skill, responsibility and mental qualities, than to those involving physical work, hazard or manipulative ability. In new plants or processes, it is common for increasing control and instrumentation to increase the points for these mental and responsibility factors, while those for physical conditions decline. It is therefore reasonable to expect higher wages for more senior operators at least. Also, since a new technique is likely to reduce the proportion of unskilled and low-wage work, the average wage will tend to increase. Thirdly, higher wages on a new plant come about simply because it *is* a new plant, because these are new jobs and because the company wish to employ good men on the plant. If these men are to be transferred from elsewhere in the company, it will be desirable to pay higher wages than they were earning elsewhere, while if they are to be hired from outside, it will be necessary to do so.

A combination of these factors therefore makes it very likely that the wage level on the new plant will be above that on the older one and this will undoubtedly ease the adjustment process for those on the new plant. But it may, of course, be that not all of those employed on the older plant can be found jobs on the new one. Even if the workers thus displaced are offered transfers to other work in the company, this other work may be less well paid than their original jobs. In this case the company is faced with the problem of easing acceptance of the change by helping employees' financial adjustment to it. In our cases, several ways of doing this emerged. In the *chemical industry*, there were two cases of protection of earnings schemes, under which employees had their previous earnings maintained for a number of weeks, and then stepped down gradually until the new rate was reached. In other cases a temporary drop in earnings occurred, though employees were quite rapidly upgraded to their previous earnings level. Another method which occurred in a few individual cases was for employees to continue at their old higher rate, but without receiving wage increases until the rate for their job caught up with them. In the *steel industry*, there were two cases of lump-sum compensation for earnings-loss. One scheme contained a

290

rather complicated formula by which workers were paid, for each year of service, a sum calculated on the basis of their present wage, previous wage and age. For a worker aged under 30, this sum was equal to the full difference between the two wages; for a worker aged 63 to 64 it was 80 per cent of the difference. The other case contained an even more complicated formula for deciding the lump sum, under which the amount paid increased with age and service, rather than decreased as in the previous case.

The general point which these examples illustrate is that management must have some kind of policy to deal with possible income-loss of transferred workers. In some of our cases, the technological changes were accompanied by a substantial change in the wage structure which made transfer into equivalent grades easier, but even in cases where workers are downgraded there may be no serious income-loss after the transitional period is over. Nevertheless, wage policy for technological change has to take the possibility of loss of income into account. This is also true of productivity agreements, for one of the main aims of these is to reduce non-essential overtime, and in the absence of other measures this would reduce weekly earnings. Hence productivity has to increase rapidly enough to allow work to be done in the shorter work-week, and to permit an increase in the hourly basic wage large enough to maintain or increase weekly earnings. Many successful productivity agreements have recognized this, and in most of them weekly earnings, though showing an increase, have risen much less than the hourly wage rate. This is the kind of case where some guarantee of income maintenance is essential if the scheme is to be agreed at all, but equally, the adjustment of the labour force to a more conventional type of technological change is likely to be easier if the company has considered this aspect of its wage policy.

(ii) *Wage Payment System*
Another major problem is that of the wage payment system as a whole. There are several questions here, and we shall begin by discussing the position of incentive payment systems in relation to technological change. The conditions under which an incentive wage system will work best are so well known as merely to need summarizing here—a standardized product, a high and steady rate of measurable output, easily controlled quality, a responsiveness of output to effort[1]—and the basic principle is that a worker's earnings depend

[1] For a full discussion, see R. Marriott, *Incentive Payment Systems* (Staples, London, 1957).

291

on the quantity of acceptable output which he produces. But on a plant with highly advanced instrumentation output is much less subject to variation due to changes in the effort of the production worker. Clearly, if workers fail to carry out their jobs properly, production will be affected, but their tasks are not manual operation, as might be the case on an old plant, but rather analysis and control. A good example is the rolling of steel sheet. In an old mill, the quality of sheet was dependent on the manual skill and judgement of the mill operator; with a computer controlled mill, output is automatically controlled, and even on a less fully automatic mill, the control of the process may be largely outside the scope of the operator. Improvements in technology have thus created conditions where payment by results schemes are much less justifiable.

In the chemical industry, a few companies in technologically advanced sectors have begun to question the basis of the hourly-paid system altogether, and at the same time, to rationalize their internal wage structure. Though these trends are not yet widespread, it is interesting to consider them, since they represent the fullest development of the effect of technology on wages. In large single-stream plants, the amount and quality of output is almost entirely outside the control of the individual operator, unless, of course he makes a mistake, or does not do his job properly. Also the manning of the plant does not vary with the level of capacity at which the plant would normally be considered operational: if it is running at all, it requires its full complement of process workers, and it is designed to operate for long periods without closing down. A system of payment related to output is, in these circumstances, unsuitable and some other method has to be used. One possibility is an efficiency bonus scheme which would relate a bonus part of earnings to efficiency in the use of utilities or raw materials and to the minimization of lost time, but the larger basic part of earnings should reflect the steady nature of production, and be calculated on a time-rate.

However, if the plant is operating continuously for many months, and the employees' workload varies only within narrow limits, it becomes unnecessary to calculate the wage on an hourly basis. To introduce the idea of the continuity of the work and the income-security which it confers, some companies have calculated employee income on a different scale from that normally adopted for wage-earners: in a recent development of this philosophy in the petroleum refining industry, Shell Oil decided to pay production workers a yearly salary at a new refinery on Teesside,[1] but the best examples

[1] See *Chemical Age*, October 7, 1967.

occur in the chemical industry, where the Manpower Utilisation and Payments Scheme (MUPS) advanced by ICI is one of the most comprehensive proposals.[1] MUPS proposed that employees should be paid a yearly salary in 52 weekly instalments, with extra payments for overtime (on the basis of a 40-hour week) and for severe or abnormal work conditions. There were to be 8 salary grades, each with a maximum and minimum level and a progression over time up to the maximum. Jobs would be distributed among these salary grades according to an assessment of three factors—mental and personality requirements, physical requirements and acquired skill and knowledge—and the MUPS agreement contained details of how these assessments would be made. The agreement signed by ICI and the trade unions at the national level had many interesting details, but some of the principles on which it was based are also in evidence in other cases.

First, the move to salaried method of payment was accompanied by a range of conditions of employment previously enjoyed only by salaried workers. These included longer holidays, generous sickness benefit and longer notice. Other agreements have provided for such 'staff status', either in whole or in part. In the electricity supply industry, negotiations during 1964 and 1965 resulted in both salaried payment and more generous sickness and holiday pay for hourly-paid workers, and a similar scheme was introduced for BEA engineering and maintenance workers in 1964. Secondly, these moves toward salaried status were normally associated with a change in the make-up of the individual's pay packet. Though earnings might not have greatly increased, a much larger proportion of earnings was made up by regular and predictable payments to workers. This is obviously so in the cases where salaried methods were used, but in the other cases as well, there was a reduction in the overtime and bonus component of earnings, and a large increase in the basic hourly wage as a result of the consolidation of bonus items.[2] This was also seen in one of our cases in the newspaper industry. Part of the workers' earnings have in the past been based on a bonus determined by the number of pages in the newspaper. This number fluctuates considerably over

[1] In July 1969 MUPS was superseded by the Weekly Staff Agreement, which contained the same kind of wage and flexibility conditions, and introduced 'no-redundancy' and 'union membership' clauses.

[2] For details see K. Jones and J. Golding, *Productivity Bargaining* (Fabian Research Series 257, London, 1966); National Board for Prices and Incomes Report No. 36, *Productivity Agreements* (Cmnd. 3311, HMSO: London, 1967); and *Productivity Bargaining*, Royal Commission on Trade Unions and Employers' Associations, Research Paper, No. 4 (HMSO: London, 1967).

the year and a recent agreement contains a formula to remove the instability this gave to earnings.

A third important aspect of the ICI proposals was that they envisaged a reconstruction of the internal wage structure of the plant, with the 8 salary grades replacing 18 previous wage grades. Again there are many examples in other companies and industries. The Shell Chemical Company reduced the number of wage grades for non-craft workers from 23 to 3 within a two-year period; Alcan rationalized 40 grades to 7; Esso Refining at Fawley formed 4 pay categories out of 18 previous grades and Esso Distribution consolidated 15 pay grades into 6. Such simplification of the wage structure will facilitate the temporary or permanent redeployment of labour between jobs. The effect may not be very marked for the more senior jobs, as such employees have had specific training, but it is important in the redeployment of semi-skilled and unskilled workers, who can then be moved more easily from job to job. Obviously the simpler the wage structure and the more workers there are in any given grade the greater is the probability that the company will be able to transfer workers to jobs with equivalent earnings, without meeting the problem of income-loss discussed above.

The MUPS scheme we have described above is obviously of considerable interest simply as a means of establishing a payment system which realistically reflects the technological characteristics of the plant. But a point of fundamental interest to the present chapter is that although the scheme was exceptionally well suited to technologically advanced plants, this was not in itself a necessary feature. A much more important condition was an increase in the efficiency with which the labour force is used. This brings us back to the problem raised in Section II of how it can be ensured that manning standards on new plants correspond reasonably closely to some optimum level, given all the difficulties which this involves. In recent years 'productivity bargaining' of the sort undertaken in connection with the MUPS scheme has offered one possible solution to this problem and we shall consider this in the following section.

IV. PRODUCTIVITY BARGAINING

We showed in Section II that there are many ways in which inefficient manning standards could arise following the introduction of new equipment. The question to which we now turn is how firms have sought to eliminate or reduce this waste of resources. For although labour costs, particularly on advanced plants, may be a small proportion of the total, they may nonetheless be the main *controllable* element. As such they will have an important effect on a company's competitive position. This indeed has frequently been the stimulus which has led firms to examine the manning standards very closely to ask whether performance in this respect could not be improved. Another stimulus has been comparison of a plant's manning with that of similar establishments in this country or overseas. For example, the Economic Development Committee for the Chemical Industry made a study of plants in North America which provided valuable comparisons with practice in the UK.[1]

But wherever the pressure for improvement comes from, it appears that before a firm can effectively tackle the problem of inefficient use of labour it is necessary for it to recognize that it cannot be solved without incurring some costs in the course of securing the benefit of greater efficiency. As an example of the opposite view one printing firm complained bitterly about the overmanning by which it was hampered, but refused to consider increasing wages in return for changes in work practices on the grounds that its workers were highly paid already. Other firms, in that and other industries, have been in a similar position but have chosen to make their workers even more highly paid rather than forgo the net savings which were realized by manning revisions. The costs to be incurred may include the wage increases necessary to secure manning concessions as well as the cost of any redundancy compensation and so on which may be necessary.

Having realized the existence of overmanning and having decided to do something about it, the firm with the easiest task is the one which has been able to retain a high degree of control over manning questions. For example, at one chemical plant a comparison of manning standards with similar plants in the United States suggested that about 25 per cent of the labour force was surplus to requirements. This was partly because of the accumulation of small technical and organizational changes which had not resulted in comparable labour force changes, and partly because of a change in the company's manning standards. Having set a target of reducing manpower by about a quarter, the plant superintendents and work-study engineers worked out a new job structure which specified the numbers and duties of the men on each plant. The next job was, of course, that of removing the surplus men, and the imaginative way in which the company approached this was discussed in Chapter 10. The company did, of course, incur certain transitional costs, in particular the

[1] National Economic Development Office, *Manpower in the Chemical Industry* (HMSO: London, 1967).

redundancy payments and the cost of retraining operators for the new occupations. But it was not necessary to give direct concessions to the union for the changes in manning, though a revision of the wage structure and a general wage increase were introduced concurrently.

But, more commonly, major rearrangements of working practices are only possible after negotiations with the trade unions, who will naturally expect some tangible benefit in return. This is the essential aspect of what has come to be known as productivity bargaining, in which 'advantages of one kind or another, such as higher wages or increased leisure, are given to workers in return for agreement on their part to accept changes in working practice or methods . . . which will lead to more efficient production'[1] As such, productivity bargaining is clearly one way of securing a more efficient level of manning on new techniques where some inefficiency has previously existed.

We saw earlier that overmanning can arise both from the policies of a single union or from union rivalries which lead to a costly lack of flexibility. But there have also been many successful productivity agreements between firms and unions to overcome such problems. For example, the introduction of mechanized publishing equipment into the national newspaper industry was discussed earlier. The first national agreement which covered this equipment, that of 1960, resulted in a high degree of overmanning in this department by firms who installed it, as it was very difficult to reduce the numbers employed even by natural wastage. But this was superseded by the 1963 agreement which, subject to safeguards, established the principle that manning could be reduced on mechanized equipment.

One company which was studied secured benefits in two ways from this later agreement. It saved money by the more efficient use of the new equipment, in line with the improved technology. But a number of other issues were also brought up in the course of negotiation which were not directly related to the manning of new equipment. For example, the number of ancillary workers such as car park attendants was reduced, enabling the remainder to benefit from a sharing of the savings. The initial purpose of the negotiations was to reduce overmanning on new equipment, but the firm was able to take advantage of the atmosphere of change to secure the elimination of other wasteful practices. Apart from greater week to week stability of earnings the main benefit to the workers was that they avoided a

1 *Productivity Bargaining*, Royal Commission on Trade Unions and Employers' Associations, Research Paper, Number 4, p. 2 (HMSO: London, 1967).

cut in earnings which might otherwise have taken place. Faster printing machinery had considerably reduced the need for overtime working, which the company intended to abolish. But by changing manning levels and sharing the savings it was possible for the working hours to be reduced without loss of pay. The union was prepared to accept revised manning on the equipment as no fundamental union principle was involved. It also appeared to accept about that time that the interests of its members in central London were better served by accepting new techniques and securing a sharing of the savings, than by concentrating on maintaining employment opportunities at their existing level.

Productivity agreements have also been used to improve efficiency by removing demarcation lines between unions and increasing the flexibility of the labour force by challenging conventional assumptions on what jobs particular groups of workers should do. The MUPS scheme discussed above was a notable example of this and provides a good example of the extent to which some companies wish to go, and the principles to which they wish to adhere.[1] It was agreed by ICI and the national unions that the optimum utilization of manpower could be achieved by the implementation, through detailed negotiations, of five principles:

(*a*) Production operators with suitable training can use tools to carry out the less skilled craft tasks which form only a subsidiary part of their work.

(*b*) In appropriate circumstances tradesmen will be expected to operate plants.

(*c*) Tradesmen and general workers can be given general supervision by men of any background.

(*d*) Tradesmen can do work of other trades which forms a subsidiary part of the main job of their own trade, according to their availability at the time.

(*e*) Support work for tradesmen can be done by tradesmen, semi-skilled or general workers as is appropriate in the circumstances.[2]

These principles were to be translated into concrete proposals by

[1] It should be pointed out here that at the time of writing (early 1969) the trials of MUPS have only just begun. Full implementation appears a long way off, and the conditional tense must therefore be used in describing the results of MUPS.

[2] *Agreement for Trials of Proposals on Manpower Utilization and Payment Structure between Imperial Chemical Industries Limited and the Trade Unions Concerned*, pp. 1–2.

negotiation on each site within several guidelines, of which three are particularly important:

(i) A job should be looked at in its totality and if it is largely craft work then it will come within the craft sphere. Similarly, if predominantly general work it will come within the general worker sphere.

(ii) That when manning jobs the aim must be to get the right man in the right job bearing in mind that men of high skill should carry out as little mundane work as is consistent with maximum efficiency.

(iii) Much retraining will be essential. . . .[1]

There are many interesting points in the ICI scheme. First, one intention was to reduce inter-craft demarcation, or reduce restrictive practices as they are commonly understood, so that craftsmen could do the work of other trades where it was an integral though minor part of a job they were doing. Presumably, therefore, a fitter doing a mechanical repair could join up simple electric wiring instead of calling in an electrician. Secondly, demarcation between craft and production workers was also to be relaxed. This type of flexibility is particularly important on large advanced chemical plants. Such plants have a very high reliability factor and are designed to run continuously between yearly overhauls. The base-load demand for maintenance is low, and the peak maintenance demand during the shut-down is very much higher.[2] Hence the traditional pattern of maintenance workers being permanently attached to a plant makes little sense, although some maintenance is still necessary on a day-to-day basis. The purpose behind the relaxation of craft-production demarcation is the evolution of 'production teams' consisting of process operators and craftsmen with their jobs merging at the lower levels. It is still intended that there should be craft jobs and production jobs, and that highly skilled men of either type should have fullest opportunity to exercise their skill, but the important point is that the MUPS scheme proposed an alternative to the traditional view of job allocations between broad groups of workers.

However, this proposed rearrangement of job allocations was not a direct consequence of technological change. Relaxation of inter-craft demarcation could occur at any stage of technological development, and might be more important in old traditional industries than in new: furthermore, the proposals were designed to apply to all of

---

[1] *Agreement for Trials of Proposals on Manpower Utilization and Payment Structure between Imperial Chemical Industries Limited and the Trade Unions Concerned*, p. 2.　　　　[2] Cf. the discussion of this at p. 271-2 above.

ICI's operations, not only to the technically progressive sections. Nevertheless, the concept of production teams of crafts and production workers really comes into its own on the new large advanced plants: these plants could be run with the traditional job allocations, but it would be inefficient and wasteful. Given this impetus to find new working methods, and having developed them, it followed that they could also be profitably used on the company's older plants.

Another example of an agreement to secure greater labour-force flexibility and more efficient manning was that negotiated by Shell Chemicals. This company agreed with the trade unions and workers that:

'craftsmen will be interchangeable within the concept of Time, Tools and Ability, on the understanding that a craftsman will retain his basic craft element and additionally will perform as required such other duties for which he is capable.'[1]

Similarly, craftsmen and operational employees would perform one another's duties within the concept of time, tools and ability, with two qualifications. First, a craftsman could only carry out the duty of senior operator under the latter's direction, and operators, when working with a craftsman, could only do craft jobs under his supervision. Secondly, though there were some craft tasks which operators could perform alone, a schedule to the agreement listed certain tasks which operators (regardless of time, tools and ability) could only perform under the supervision of a craftsman.

The Shell Chemicals agreement raises an interesting problem, namely how to progress towards this goal of interchangeability, and here there are two possible approaches. The first is to define certain jobs, previously the preserve of one group of workers, which may now be done by other types of worker: this was the method used by Esso at both Fawley and Milford Haven, by BP, and Mobil Oil at their Coryton refinery. The second is to agree on the general principle of flexibility, and to allow specific proposals to evolve. As the Prices and Incomes Board pointed out:

'both . . . have their dangers. . . . With a general statement of intent the danger is that nothing will be changed; but to enter negotiations with a "shopping list" of restrictive practices may lead to the setting of a separate price on each item. The result could be an agreement by the unions to renounce, for a considerable price, only those practices

---

[1] *An Agreement between Petrochemicals Ltd. and the Trade Unions*, p. 4.

which are by themselves insufficient to permit major changes in methods of operation.'[1]

The ICI scheme, as agreed with the national unions, was a statement of general principles, and the agreements on the application of these principles by detailed negotiations at the site level would have devised precise proposals. The Shell scheme appears to have followed a different course and to have advanced further along the road to complete flexibility, within the concept of 'time, tools and ability'. There, the principle of flexibility was agreed, and this was held to mean interchangeability except for certain listed jobs that a process worker could *not* do alone. Outside this prohibited list, flexibility operated.

Examples of demarcation problems which have proved much less tractable can be drawn from the steel and printing industries. We have already discussed the controversy between the AUBTW and ISTC over the lining of LD converters. A rather similar dispute was that between ISTC and the Amalgamated Engineering Union[2] over the manning of a new machining operation, which was claimed by both unions. In the printing industry there has for some time been a serious conflict between the craft NGA and the non-craft SOGAT over the boundary between their respective spheres of jurisdiction.[3] Each of the cases mentioned has at some time proved to be beyond the ability of the normal collective bargaining machinery of the industries concerned and resort has been made to arbitration or a Court of Inquiry. This has always settled the immediate dispute, but has not necessarily solved the problem as a whole—the dispute between the Confederation and the bricklayers appears to have arisen in at least four different plants and the Court of Inquiry into the dispute between the NGA and SOGAT has not removed the fundamental grievance of the latter.

This suggests a clear limitation to the effectiveness of productivity bargaining. Many inefficient work rules have grown up in a fairly casual way at the place of work and involve no particular matter of principle on the part of the union concerned. Productivity bargaining appears to be a very useful procedural device for negotiating the removal of practices of this sort. But it may be much less successful where demarcation issues are involved. These may represent the unions' efforts to retain the craft identity in the face of changing conditions, and as Flanders points out, fundamental beliefs of this

1 National Board for Prices and Incomes, *Report* No. 36, op. cit., p. 6.
2 Cf. Chapter 7.
3 Cf. Chapter 4, Section IV.

sort may be much less susceptible to modification by argument or inducement. Even if the men directly involved would themselves gain financially while retaining their own security with the firm, there may still be resistance by union officials or shop stewards to proposals which threaten, however remotely, the separate identity of the craft, while conversely, an agreement may be concluded by the trade union only to be subsequently rejected by the membership. Thus, at Fawley, the Blue Book proposals on inter-craft flexibility were substantially scaled down in negotiations, while the further relaxations proposed in the Orange Book ran into intense opposition which was only partly overcome. On the other hand, a union's view of the subjects about which it will or will not negotiate are not fixed. For example, the trade unions have made concessions in some companies which they would not have been prepared to make to other companies or in industry generally, presumably reflecting a good history or environment of industrial relations as well as the secure future prospects of a particular company.

However, a more fundamental problem remains, and that is how far productivity bargaining can be used in determining the manning of new equipment when it is installed. We have so far discussed it in cases where, some time after a new technique had been installed, it was decided that overmanning had developed and the firms sought its removal through some form of productivity agreement. Similarly, one can envisage that as a new plant is installed, it may be beneficial if certain grades of work are amalgamated to remove distinctions which are no longer relevant, so securing more efficient working on the new plant. Again, this may be more easily achieved through a productivity agreement which gives the workers some benefit in return for their departure from traditional practice.

But to what extent should the general manning levels on the new plant be the subject of productivity bargaining? Although changes in working practice may make a contribution, the main source of increased efficiency is the investment in capital equipment embodying new techniques of production. In this case it may be necessary for the greater part of the savings to go to the firm to give a satisfactory return on the investment, and consumers will also expect to benefit in the form of more stable prices. Most of the firms in our study gave the workers involved in a technological change some increase in wages, but these should be regarded as a reflection of the firms' general wage policy rather than as a productivity bargain in the true sense.

But whether manning standards and wage questions are settled

301

within the framework of rigorously defined productivity agreements, or in some other way, there remains the issue of whether these problems are best solved at the national or plant level. National agreements on manning scales seem to be rather rare, though they do exist. For example, in the printing industry, there is a national agreement between the BFMP and the NGA which lays down the manning of certain types of printing machine. In steel the only national agreement on manning has been that introduced to cover the operation of LD converters. This is what one would expect, in that national agreements are only useful in industries operating with quite a lot of fairly standard equipment, such as the railways or engineering. National negotiations may also be useful for establishing rules governing the ownership of work by a particular group of workers. For example, in printing there is a series of formal and informal agreements between the various unions on the dividing lines between their respective functions and these are almost universally followed by firms when filling a job. Further, the distinction between maintenance craftsmen's work and other work is generally accepted throughout industry, either formally or informally.

But where plants are large and complex, as for example in the chemical industry, national agreements on manning are much less useful. Further, many types of equipment are unique to a particular plant, and even where standard machines are employed the environment in which they are used or the demands made upon them vary substantially from plant to plant. Similarly, problems such as the rate at which machines run or the extent of ancillary assistance, both of which affect the manning of a plant, are very much matters for deciding in the light of circumstances at a particular firm.

Rather similar considerations apply in the case of wages, in that the extent to which national bargaining mechanisms can solve the wage questions arising from technological change is rather limited. A possible case could, for example, be in an industry which was relatively small and homogeneous and where particular technological changes were being rapidly diffused throughout the industry: there it might be possible at the industry level to reach agreement on the wage policy to be adopted by individual firms when the new technique was introduced. But, in general, as with manning, it is difficult to take account of all possible local circumstances in bargaining at the industry level, and, therefore, if national bargaining is to help with the wage problems of technological change, it will best do so only by providing a framework for discussions at the local level.

Obviously, protection of earnings schemes must be designed to cope with the conditions of the individual case, and so must wage changes leading to anticipated redeployment and flexibility, while wage increases based on productivity require accurate estimates of past productivity trends, preferably at plant level. But the relationship between industry and local or plant bargaining is most clearly seen from productivity bargaining. This can begin at the national level and be approved by employers and unions at that level, as was the case with the ICI and British Oxygen schemes, which were company agreements negotiated at national level, and the electricity supply agreement among others.[1] On a wider basis, company productivity bargaining can be encouraged and assisted by employers' associations which have sufficiently flexible rules. For example, the Chemical Industries' Association and the trade unions have recommended guide-lines on productivity bargaining and have set up a working party to advise those who are attempting to conclude a productivity agreement. But the detailed negotiation of changes in work practices and the necessary wage changes associated with them must inevitably be carried out at plant level, and the particular solutions reached will be specific to that plant. If national agreements prevent this kind of negotiation and forbid the large wage increases which might be necessary, they cannot in any way set up a framework for plant productivity bargaining. But, of course, freedom from possible restrictions caused by industry-wide national bargaining is only the first step. There must be decentralized negotiation on the details of the agreement, and the most difficult phase of productivity bargaining has been the development of effective plant negotiations to secure the agreement of the workers to the changes in work practices and the wage increases to be paid.

The spread of plant negotiations of this kind would have far-reaching implications for the structure of trade unions and the whole system of collective bargaining and industrial relations,[2] but it is obviously possible for many of the manning and wage problems of technological change to be solved without this degree of formality. The point is, though, that the manning and wage problems resulting from technological change and the adoption of appropriate policies towards them are as much an adjustment problem as are the problems of transfer, training and redundancy to which most attention is given. They must, in the same way, be forecast and planned for by the employer, if successful adjustment to the change is to be made.

[1] See National Board for Prices and Incomes, *Report* No. 36, op. cit., pp. 35-7.
[2] This idea, of course, underlies much of the argument of the Donovan Report.

303

CHAPTER 12

REDUNDANCY, RETRAINING AND RESETTLEMENT

I. INTRODUCTION

In Chapter 1 job redundancy—the disappearance of a certain number of jobs—was distinguished from worker redundancy, the actual release from employment of men because jobs have disappeared. This distinction can be usefully introduced again here. When a company announces that a plant or part of a plant is to close, the number of job redundancies and the number of potential worker redundancies is equal to the employment at the date of announcement. Faced with this fall in labour requirements because of the effect of technological change, a company is likely to have recourse first to a series of measures which will allow a reduction of the labour force while the unit remains in operation. These 'early adjustment devices' include natural wastage accompanied by a stop on recruitment wherever possible, and early retirement for employees near retiral age. Natural wastage is perhaps the most important method of cutting the labour force, especially when it can be accompanied by cessation of activities not essential to the continued long-term operation of the unit. For example, the company may be able to close certain departments or have work done by outside firms on a short-term contract basis, so allowing its own employees to leave. In so far as these early adjustment mechanisms help to reduce the labour force during the period before the closure, the number of job redundancies is reduced and so is the number of potential worker redundancies. Once full advantage has been taken of these mechanisms, the company is faced with a choice either of employing some other adjustment devices under which the worker would stay in its employment, or declaring him redundant. The most important of these other adjustment devices are transfer or redeployment of workers to new jobs which are similar to those which have disappeared, transfer to different jobs for which retraining is necessary, and resettlement under which employees are offered the opportunity of employment at a plant beyond daily travelling distance of the affected plant. This chapter deals with the extent to which redundancy, redeployment and resettlement are alternative courses of action for a company faced with a local labour-force surplus, with the experience of our industries, and with the more general problems of labour-force adjustment.

In some situations the employer has no choice as to which adjustment mechanism to use. If a company has no other employment available anywhere, as for example when a firm closes down, transfer and resettlement are not possible, and all employees remaining at the time of closure must be declared redundant. Also, transfer is only feasible when a firm has employment opportunities in the vicinity of the affected plant, and resettlement can only occur when it has vacancies in another plant outside daily travelling distance. Transfer and resettlement are thus often not alternatives to redundancy or to one another, but the most complex decision by the employer occurs when he has the choice of all three adjustment devices, and it is useful to consider this particular case. It is also helpful to outline initially the way in which the employer might make a purely economic analysis of the situation. If he were to act solely on economic criteria, the employer's decision would be determined by the relative costs and returns of transfer with and without retraining, of resettlement and of redundancy. The factors determining these costs and returns would vary between situations but the information which would ideally be required would include the following:

(a) the similarity of any vacant jobs to the jobs of existing potentially redundant workers; the more similar the jobs, the more likely is transfer;

(b) the nature and cost of retraining these workers and their existing skill level; the most difficult and costly the retraining, the more will the employer consider the possibility of new recruits;

(c) the level and structure of unemployment; this will enable him to estimate the availability of labour;

(d) the levels of wages and earnings in the company relative to those of competitors in the labour market;

(e) the prospective level of wastage of new recruits; this will influence the cost of recruitment;

(f) the level of redundancy pay due to existing employees;

(g) the possible cost of resettlement, e.g. movement costs, housing allowances, compensation, etc., and its likely long-term success;

(h) any reason for different levels of productivity between new and existing workers.

Where the company has employment both in the area of the affected plant and elsewhere, the information listed above would be required for several groups of employees in more than one labour market. It would be possible to analyse the employer's alternatives much more exhaustively, but enough has been said to indicate the complexity of an economic assessment of the employer's choice. If we wish to establish from this a hypothesis as to how employers are liable to act in an actual situation of labour-force surplus, three qualifications must be introduced. First, institutional or contractual restraints on employer action may exist. Secondly, an employer may not choose or may not be able to make a purely economic assessment because of social considerations. Thirdly, we cannot assume that the employees themselves are passive agents in any redeployment or resettlement, and that the employer can control the process and choose the course of action which best suits him.

Institutional restraints on employer action can come from various sources. Although a company may be able to determine unilaterally how potential redundancy is to be treated, in a specific redundancy situation it will inevitably be subject to trade union pressure on how the particular redundancy should be handled, not only on the amount of payment to redundant workers, but more commonly on the policy used to determine the speed of the labour-force cut-back and to decide which employees are eventually to be declared redundant. This may be done through the seniority principle which favours the continued re-employment of long-service workers, and may limit an employer's freedom to retain or transfer individual workers, so that workers whom it would be most efficient to re-employ cannot be retained in the most suitable jobs. Other types of institutional control affect the deployment of labour more directly. Negotiated work rules on plant manning may limit the employer's control of redeployment, and the possibility of retraining existing employees is circumscribed by apprenticeship regulations which specify the duration of training and place limits on entry. Further problems may come from the unwillingness of some craft trade unions to accept even retrained craftsmen and the general non-acceptance by the relevant unions of semi-skilled workers retrained as craftsmen. These limitations throw up a number of institutionally created non-competing groups which inhibit efficient redeployment. There may also be legal restrictions on employer action, but these are normally concerned only with the period of notice to which employees are due and with the amount they must be paid under the Redundancy Payments Act. This Act and other manpower measures may have more effect on the employer through their influence on the costs and benefits of various courses of action.

So far, we have assumed that the costs and returns of employer action were calculable, but they are obviously extremely difficult to assess. For example, the cost of retraining existing workers as against the cost of training new workers may be difficult to determine even after the event, and the forecasting of the training costs, of success rates, and of future productivity must be subject to wide margins of error. Then, when considering the desirability of hiring new workers rather than transferring existing workers, the employer must make an estimate of the likely cost of recruitment, taking into account likely wastage as well as hiring. Few employers in Britain have any precise data on the *actual* cost of recruitment, let alone the estimated cost. Another cost which is more difficult to establish than it might seem is that of resettlement. While the cost of transfer grants, moving allowances and resettlement compensation can be published and advertised to employees, it is difficult to forecast the number of employees who are actually willing to move, as distinct from those who profess an interest in moving. In this instance the reaction of existing and prospective employees is difficult to forecast, and in a general sense the reaction of employees to employer policy is one of the essential qualifications to the economic evaluation. While the employer has three courses of action—transfer with or without retraining, resettlement, or redundancy—he cannot juggle around with the workers as he thinks fit, for they have a choice as to whether or not they will stay with the firm or fit in with the course of action the employer decides on.[1] In so far as labour mobility can be viewed as an economic process, the employee will also make an explicit or implicit calculation of the costs and returns of movement to a new job, a new occupation or a new area, and the most satisfactory course for him may not be the best from the employer's point of view.

It is not only the unpredictable nature of the employee response which may influence a company's reaction to a redundancy situation. Employers may feel a social responsibility towards their workers and may be prepared to make an attempt to re-employ them even if this would not on the face of it seem to be the most economic course of action. Similarly, the level of voluntary redundancy payments and the policy adopted towards the planning and timing of manpower changes

[1] In some cases where redundancy is almost bound to be the course of action (e.g. a complete closure), it may not seem to matter whether employees leave. But if the workers' time-table for leaving does not fit the employer's, problems may arise.

can reflect the social obligation which the employer feels is placed on him to act in a responsible way. The choice between redundancy, resettlement, retraining, and recruitment may therefore be made primarily on social grounds, though it is unlikely that economic factors will be entirely ignored. For example, it would be futile to attempt to retrain unskilled workers as fully skilled if the prospects of success were negligible and if workers of that skill were readily available on the labour market, or if institutional restraints were to prevent the utilization of the retrained workers. But where the economic factors are well balanced or where information to make an economic choice is not readily available—and this may usually be the case—social factors may be important or pre-eminent. This is of particular importance since many will consider that the real cost of redundancy or transfer is the hardship or disruption of life suffered by the workers concerned. Our study was concerned with company policies, and did not directly cover this aspect of redundancy, though some attempt will be made to assess the adequacy of redundancy schemes in our cases. The next section examines the evidence of our cases on redundancy, retraining and resettlement, while Section III considers more generally the problems of adjustment to change and the role of public policy.

II. THE EXPERIENCE OF OUR INDUSTRIES

In our three industries we examined altogether about twenty cases where technological change had had a marked effect on the labour force, and most provide some illustration of company policy towards transfer, resettlement or redundancy. In order to appreciate the choice facing the employer in particular labour-surplus situations, it is necessary to examine cases individually, and the Section therefore summarizes by industry the cases in which transfer, resettlement and redundancy were important, and the importance of social or institutional factors in the employer's choice of action.

(i) *The Chemical Industry*
In the chemical industry seven cases had transfer, resettlement and redundancy as important features.

(*a*) In the largest closure and redeployment, 146 and 245 hourly-paid employees on a large chemical site were transferred in two separate exercises after wastage and early retirement had reduced the number to be redeployed by about 25 per cent and 30 per cent respectively. The transfers were accomplished through a system of internal publication of vacancies and a deliberate policy of moving potentially redundant workers to suitable jobs. For process workers, retraining involved at most on-the-job training and familiarization; for maintenance workers most redeployment was also to comparable jobs, though some maintenance workers were given company instruction in new techniques relating to their craft. In addition, about 20 skilled fitters were retrained as instrument mechanics. For employees transferred to less well-paid jobs, a protection of earnings scheme ensured that they received their previous level of earnings for a month and then had pay stepped down gradually. Redundancy was regarded as a last resort, and virtually all workers who did not retire or leave voluntarily were redeployed. The demand for labour in the local area was high and it was therefore in the employer's interest to transfer suitable workers. It was admitted, though, that some men were redeployed to plants which did not really need them at that time, and that the continued employment of about 20 or 30 men was largely a social obligation and had no economic justification.

(*b*) In another similar closure involving 170 workers, an attempt was again made to redeploy all workers. The demand for maintenance workers ensured ready transfer into comparable jobs, but to assist in the redeployment of process workers, a special training scheme was set up with retraining courses of a week's duration to help the process of on-the-job training operate more quickly. Protection of earnings was again given, with a week's protection per year of service before scaling down. Some older workers who proved difficult to redeploy were 'carried' for a time as reliefs, odd-job men, etc. Seniority was important in this case, and long-service workers from the closing plant had the right to 'bump' certain shorter-service men from the plants to which redeployment was proceeding. This altered the pattern of redeployment but did not cause redundancies: as in the previous case, management chose to avoid redundancy whenever possible.

(*c*) A smaller closure on another large site again used a system of internal advertisement by which potentially redundant workers could apply for vacant jobs. About 100 workers were originally affected, but wastage was very high and few workers had applied for transfer. By the end only 14 had done so. The jobs were largely unskilled and physical: few workers were capable of being retrained to a higher standard, and most were not offered or did not want transfer. Consequently there were redundancies here, but the local demand for labour was very high, and when the closure finally came there were many temporary employees.

(*d*) In this case, transfer was not possible since the company had no other employment in the area. To reduce the number of worker redundancies a voluntary redundancy scheme was introduced. An employee of over 52 years of age could ask for release at any time, and an employee under 52 could do so if the notice of his redundancy had been given. If he could be spared or replaced with a temporary worker, he could take redundancy payments and leave: 62 employees did so. 107 hourly-paid workers accepted resettlement to other parts of the country. The company's financial assistance included moving expenses, a subsistence allowance to cover extra expenses or a lower wage, help with housing directly or through a local authority, and travel-back allowances to facilitate contact with the home area. These allowances ceased after six months. Though it had no local employment available, the company offered retraining to workers to help fit them for another job. 143 requests for training were received, mainly for vehicle driving, clerical work, and semi-skilled work, but many were impossible because of lack of ability or facilities. 20 craftsmen received refresher courses in aspects of their craft. 125 employees were eventually made redundant. In this case, some redundancy was inevitable, but the voluntary redundancy scheme and the success of resettlement meant that the number declared redundant was lower than might have been expected.

(e) Two other cases of closure of isolated plants did not show resettlement in the same light. Few employees wished to be redeployed to other plants even though the distances were relatively short, less than 25 miles. In one case the jobs to which the employees were offered resettlement were quite dissimilar and considerable retraining would have been involved, and no manning problems were experienced at the plants to which the workers refused to go. There was considerable redundancy in these two cases; early retirement and the premature voluntary release of non-essential employees helped to reduce the number eventually redundant, but only 3 employees in one of the cases actually accepted resettlement. In one case the redundancy did not have severe effects since the demand for labour in the local area was very high, in the other, job prospects were more difficult. However, this occurred after the Redundancy Payments Act and earnings-related unemployment benefit had been introduced and the financial terms of the redundancy were very favourable.

(f) One case of manning-up presented transfer and redundancy not as alternatives, but successively. Workers had been trained on existing plants most similar to the new units to allow them to gain experience of operation of the broad type—the so-called 'double-manned' method of training. Following the transfer of the most suitable workers to the new plants where further on-the-job training was given, the company decided to revise its manning, and instituted a voluntary redundancy scheme.[1] The most notable factor of this was that the company decided to use no selection principle whatsoever in the course of the redundancy. Though the extent of the over-manning was estimated by trade and grade and the numbers of each type of worker allowed to go were limited to this excess, within these limits any worker was free to volunteer for redundancy, even if he was a worker whom the firm would wish to have kept. The company's stated objective was to accomplish the cut-back with the minimum personal hardship to employees, and it was prepared to allow those workers to go who had no particular ties with the company. Alto-gether 368 men left the company under this scheme, at an average cost to the company of about £250 per man.[2]

The chemical industry cases lead one to the conclusion that em-ployers were unwilling to consider redundancy as an alternative to transfer or resettlement unless it was impossible to avoid it. Every attempt was made to transfer employees into other jobs, and because of the large sites on which many of the cases occurred, this was possible even when a complete production unit was closed. In several cases, companies re-employed workers who were not in the economic sense essential, and offered resettlement to workers who might not have been as efficient or productive as new recruits in the reception area. The experience of these cases was that retraining was in general no problem. Process workers normally required only on-the-job training, perhaps supplemented by short formal courses in the case of plant operators, and companies were largely successful in redeploy-ing workers into comparable jobs. Maintenance workers too could normally be fitted into similar jobs to those they had left.

There was little evidence that the employers' choice of adjustment mechanism was dictated by trade unions, or that the unions had very much effect in limiting the employers' freedom of action. This was no doubt partly because of the lack of retraining problems: in only one case did retraining of workers for work of craft status occur, and since these workers were already craftsmen no acceptance problem oc-curred, though it would have done had the company attempted to introduce upgraded process workers. Seniority rules were important in only two cases, and this was in determining to which jobs potentially

1 Further details can be found in Chapter 10.
2 This scheme pre-dated the Redundancy Payments Act, so the whole cost fell on the company.

redundant employees could be redeployed. The reasons for the apparent lack of trade union reaction were the success of the transfer policy, which meant that there was little redundancy, the flexibility of company policy, and the extent to which the companies were prepared to minimize hardship to workers affected by technological change.

The importance which companies attached to this can be seen from the policies followed. The period of notice of closure was typically 12 to 18 months, and during this period all companies used early retirement and natural wastage to assist in the run-down. As closure approached, voluntary separation without loss of redundancy compensation was a feature of several cases, as was the hiring of temporary workers as replacements. Redundancy schemes were in existence before the cases occurred though special provisions were arranged in some instances, and the levels of compensation were in advance of the normal practice of the time. Perhaps the best examples of how social responsibilities affected company policy and increased the direct cost of adjustment, were the retraining schemes offered for outside employment in (d) above, and the voluntary redundancy scheme in (f). Section I described employer choice in a labour surplus situation, but it is hardly an exaggeration to say that in our chemical industry cases the choice often lay rather with the *employee*: transfer or resettlement were offered as alternatives to redundancy, and while transfers were mainly accepted resettlement was not.

(ii) *The Steel Industry*

In the steel industry, there are four cases of relevance to this part of the discussion: all cases were prior to the introduction of the Redundancy Payment Act.

(a) In 1962 the closure of a small works, due to the concentration of activity on new larger-scale plant in other works of the same company, led to the disappearance of 70 staff and about 320 hourly-paid jobs. The company arranged that at its other local plants preference would be given to the workers who were to be made redundant. Details of age, service, job history and earnings were circulated to these other works, and by the time of the closure all staff workers had been found new jobs there without suffering a loss of a salary. Of the hourly-paid workers 25 retired and apprentices were transferred to other branches. Of the remainder, all those under age 60 were offered alternative work in other plants, about 280 in total. 169 men did not accept and became entitled to either severance payment (a straightforward redundancy compensation payment) or resettlement allowance (a supplement to unemployment benefit paid while the worker remained out of work). The average payment was about £63. So far as can be determined, these were all workers with at least five years' service, for whom the prospect of redundancy compensation appears to have outweighed the availability of alternative employment. There also seems to have been a high degree of attachment to a narrowly defined local area. Transfer of hourly-paid labour (other than apprentices) was thus accepted by about one-third of the total works labour force, to works within a fifteen-mile radius or less. The men transferred in general moved to comparable work and no major retraining programme was involved.

(b) A larger closure[1] in 1964 led to a loss of 162 staff and 853 hourly-paid jobs. Because of the size of the closure and other developing employment opportunities, the closure was phased over a period of several months. A favourable local labour market situation meant that a relatively high proportion of workers were able to find alternative work without assistance from the two major companies. However, considerable efforts were made to offer alternative employment within the company organizations, and approaches were also made to other local employers, some within the steel industry. 126 members of staff were found new jobs, mainly within the organizations of the two companies. 44 apprentices and junior operatives were transferred to other works in the area, though 4 remained out of work. One woman out of 40 was found another job. Of the remaining 760 or so hourly-paid employees, 385 were re-employed within the steel industry as a result of the programme, while another 50 were found jobs in other local firms. Three months after the final closure 80 remained unemployed. It would seem, therefore, that about 250 found work on their own initiative. By far the greatest amount of re-employment took place within the local area (roughly within 12 miles of the plant that was closing), and many resettlement offers were made, both within the companies concerned and elsewhere, but were not taken up: of 250 workers who were offered jobs in Scunthorpe with resettlement allowances, only 30 accepted. Redundancy payments were made to over 200 workers and staff, and 48 of these also received a supplementary benefit while they remained unemployed.[2] The minimum period of service for eligibility was two years. Hourly-paid workers were given what amounted to five weeks' notice of redundancy, and any employee for whom work was not available

---

[1] This is the Baker-Bessemer case, referred to in Chapter 7 above, where two large companies were responsible for the closure of a smaller company.

[2] The average payment for staff was about £500, with 7 receiving over £1,000. For hourly-paid employees, the average was just under £200.

during that period received his average weekly wage of the previous 12 weeks.

(c) The third case is the SPEAR programme discussed in Chapter 7 above. The construction of a new electric melting shop to replace two existing shops led to a disappearance of just over 900 jobs. A major effort was made to avoid worker redundancy, and internal redeployment, retraining and transfer all played a part in the final outcome. It was decided that existing promotion lines should be preserved and that the best men available should be selected for the key jobs on the new plant. Jobs becoming available in other parts of the works were notified to those not so selected and only when these jobs were not taken up by existing personnel were new workers recruited. Even then new recruits were hired on an explicitly temporary basis and normal wastage was not replaced except where absolutely necessary. Attempts were also made to transfer workers to other works in the company organization but the resettlement offers were accepted by only 34 men, despite company efforts to help with housing and movement problems. A further 9 men were placed in other companies. Retirement and death accounted for about 190 workers. The main burden fell on internal transfer, amounting in total to 544 men, many of whom were retrained by the company. Retraining was most important for the 72 melters in the new shop, who had to become proficient in a new steel-making practice. Two special courses—one theoretical, one practical—were run for these workers, and those selected had to pass certain tests. The first team to take over one of the new furnaces also had experience at another works. A 12-week retraining course was also provided for shift managers and controllers, and maintenance craftsmen were given a much shorter course of instruction on the new plant. For the others, the problem was mainly one of job familiarization. Redundancy was confined entirely to the temporary workers. Of the 500 employed in this way at the end of the programme, 250 were made permanent and about 140 paid off. Those on a temporary basis were not eligible for redundancy payments, but the company did have an extensive compensation scheme for permanent workers who suffered a reduction in earnings because of change of jobs.

(d) In the fourth case, involving the replacement of an old rolling mill, there was again a reduction in the number of suitable jobs, and 151 men from the old mill had no comparable job on the new mill. Of these 40 were found jobs on the new plant, and another 43[1] left

1 Almost all of these were under 40 years of age, one-half had less than 5 years' service, and three-quarters lived in towns more than 10 miles from the works.

the company voluntarily during the transition period, which was planned to be phased over a 6-month period (though in fact the period was extended due to an increase in demand). The remainder were redeployed to other parts of the works, a process which was helped by weekly and sometimes daily bulletins on jobs becoming vacant within the works, and a comprehensive programme of interviewing and personal assessment. 31 of these redeployed workers were eligible for compensation since their new jobs in the works carried considerably less pay than the old. The average payment was just under £90 (this was in 1961). Retraining was not a serious problem, though a short course was arranged for key workers (rollers, drivers, furnacemen, etc.) and groups were given short spells at other works in the company where similar plant was already in operation. While the plant was being constructed it was possible for the key workers to familiarize themselves with the physical lay-out of the new mill

The experience of these four cases, and more fragmentary information concerning other cases in the steel industry, suggests a number of conclusions. Prior to the Redundancy Payments Act the companies appear to have tried, wherever possible, to avoid worker redundancy and even in works closures where some redundancy was inevitable, efforts were made to provide re-employment by means of transfer to other works in the same organization, to other companies in the same industry or to other industries in the same general labour market area. Where internal redeployment has been important, there seems little doubt that the companies have put their social obligations above the purely economic considerations, though as in the chemical industry cases local labour shortages may have been an added encouragement to firms to adopt a policy of retention.

In only one case was retraining a major problem and here very considerable efforts were put into selection, methods of training and instruction in practical operation. The seniority rule affecting the promotion of production workers in cases of redeployment was perhaps a more limited influence than might have been expected, since existing promotion lines were preserved (i.e. melters for a new plant would be selected from those already in melting teams, crane drivers from the existing pool of crane drivers, etc.) and the selection of individuals from within these lines was reasonably flexible. In general, the trade unions accepted the need for technological change and its implications: early consultation by companies with the relevant unions was probably of considerable help in this respect, and while there were disputes, these were few in relation to the

315

size of the capital investment programme in a period of changing technology.

In all cases the companies had prior arrangements to deal with redundancy or compensation payments, but changes were made in these schemes on an *ad hoc* basis to meet the needs of the particular case or to bring the scale of payments up to a level more in conformity with the times. Perhaps the most striking features of these arrangements was the willingness of companies to compensate workers who were retained in the company but who suffered a reduction in levels of earnings. Some companies also paid a supplementary unemployment benefit to workers who were made redundant and who could not at once find new jobs.

As one might expect, unit closures presented a more serious problem than other types of closure, and in part this may be due to the periods of notice the companies were able to give. In cases (*a*) and (*b*) above the closures were announced fairly suddenly, and the maximum period over which the rundown was planned was about six months. The evidence from the chemical industry showed that closures were often planned over a longer period and employees given longer notice. This may partly be due to the types of closure: in the chemical cases there was usually a new plant opening up to replace the old, and once the commitment to build the new plant had been taken the decision to close the old plant could also be made.[1] In cases (*a*) and (*b*) in the steel industry, there was no new plant opening up, though production was being transferred to more efficient and modern plant. One explanation of the sudden closures may be that the output of these high-cost plants is sensitive to the swings of the trade cycle, and the decision to effect closure has to be taken quickly if losses are not to build up. However, the period of notice which was given, though short by the standards of the chemical industry cases, nevertheless seemed to be long enough to allow successful adjustment of the labour force.

(iii) *The Printing Industry*
Four main cases in the printing industry will be summarized briefly, with the experience of other smaller cases being taken into account in the general discussion.

(*a*) The largest closure was that of an old general printing factory in central London, as part of a widespread reorganization of a group's

[1] As Chapter 10 showed, the precise timing of the closing was much less certain, and two potential cases of closure disappeared when an upsurge in demand led to a reprieve for the old plants.

printing facilities. In anticipation of the displacement problems that might arise the company had drawn up a policy to deal with them. The main element was to introduce a fixed retirement age and improved pension scheme, neither of which had previously existed, and strict control was to be exercised over new entrants to those areas of the group's labour force where surpluses were expected to arise. The policy also intended to find alternative employment within the group's London factories for those of working age displaced with transfer or resettlement payment being made and pension rights being preserved. These measures, which it was hoped would minimize the amount of redundancy, came into operation in November 1963. The first major closure, at Odhams, came in October 1964 and resulted in 1,218 workers being displaced. 748 workers were transferred or resettled to the various London factories of the group, mostly to comparable work. A partial exception to this were the 164 men who went directly to a new plant at Southwark Offset which also recruited workers from other group factories, their vacancies being in turn filled from Odhams. A comparison of the occupations of 374 employees can be made with their earlier jobs. Many, particularly in binding occupations, moved to the same jobs, but others changed occupation. For example, keyboard operators were engaged on a different type of machinery from those earlier operated. There did not appear to be any serious disadvantage to the company in having its choice of recruits for Southwark Offset limited to present employees, for the firm was able to recruit from any group factory, not just from those displaced from Odhams, so that the choice of labour was relatively wide. Those transferred between factories received a tax-free payment of four weeks' basic wage and the cost of living allowance, and in addition one week's payment for each year of service. 470 workers left the group's employment. 200 of these retired, reflecting company policy to concentrate the reduction on those above normal working age, 100 left voluntarily before the closure, and 170 were declared redundant. The latter received a tax-free payment of four weeks' basic wage plus redundancy pay of two weeks' basic wage per year of service.

(b) In 1963 a large evening newspaper was merged with its competitor and production concentrated at the latter's plant, which was able to produce the combined paper with only a negligible increase in staff. Some 444 production staff were displaced, but the problem was largely overcome by the fact that these workers were employed in a plant which printed certain editions of several national newspapers on a contract basis. It was thus possible to transfer many of those not

317

required for the evening papers to other work during the day or at night. Eventually 157 were transferred to night staff, 213 remained on day-work, 53 retired and 91 were declared redundant. The latter were men who were offered alternative jobs on day or night work but refused them, and they received redundancy pay on the following scale:

1–4 years' service—1 week's basic wages per year of service.

5–14 years' service—2 weeks' basic wages per year of service.

15 or more years' service—3 weeks' basic wages per year of service.

As was quite common in newspaper closures before the Redundancy Payments Act, no advance warning was given, the announcement of closure being given on the day of the last issue. The firm recognized the difficulties this might cause and paid wages for the two weeks following the closure, though the men were not required to report for work. Redundancy payments were made on the basis of length of service to all employees who were not re-employed.

(c) A company which installed automatic stacking equipment in its publishing room found that this would reduce the labour required from 159 to 121, a saving of 38 semi-skilled men. Under the terms of an agreement with the trade union concerned, it was not able to declare the men redundant, and the numbers could only be reduced by natural wastage or transfers to other departments. The prospects of such transfers are somewhat limited in printing by the trade union structure, and also by the relatively low rate of turnover in the industry as a whole. In view of this, the company allowed numbers to fall by not replacing such natural wastage as did occur during the eighteen months prior to the new equipment being installed. It also transferred workers out of the department as opportunities elsewhere arose. In this case, then, the only alternative open to the company was that of internal transfer of the workers who did not leave voluntarily.

(d) Finally, we may illustrate the very substantial reduction in labour requirements following the introduction of mechanized publishing in national newspapers, by examining the experience of one firm. With the installation of mechanized publishing equipment, the labour force required for a particular type of newspaper fell from 688 to 533. The labour force had previously been made up of a regular basic staff supplemented on a night-to-night basis by casual labour. The agreement with the union did not permit any regular employees to be made redundant, and again, internal transfer was impossible because of the union structure. The solution was simply that the firm ceased to call in any casual workers from the pool, and at the same time arranged for the employment of regular staff sufficient for its requirements. This case illustrates a situation where regular employees were protected from redundancy, the brunt of the loss of job opportunities falling upon the casual element in the labour force. To reduce the supply of this type of labour the union had prevented any new entrants to the branch, and had prohibited rest-day working.

The last two cases in particular illustrate an important feature of the adjustment process in printing, where redundancy has been rather rare. These are the 'no-redundancy' agreements which have been negotiated either nationally, as in the case of the Process Trade agreement of 1960 and the Mechanized Publishing agreement of the same year, or at plant level, as in case (c) above. This can clearly lead to under-employment: for example, a newspaper employing 10 process engravers installed powderless etching equipment which reduced its labour requirements to 7 men. Because of the no-redundancy agreement, it continued to employ the men, and even when one man left, pressure was put on the company by the union until it replaced him.

How serious this problem is depends of course on the ease of internal transfer. Internal transfer may be affected either by institutional obstacles or by the cost and feasibility of retraining. In none of our cases were displaced non-craftsmen transferred to craftsmen's jobs or vice versa, but craftsmen of one union have been transferred to work generally done by craftsmen of another union. For example letterpress machine minders (members of the NGA) have been transferred to work on web-offset machines, normally manned by the ASLP, and the latter union has allowed displaced machine minders (NGA) or stereotypers to be employed in making offset printing plates. Generally training problems have not been very important, partly because a feature of technological change in printing has been the 'de-skilling' of certain jobs, such as offset plate-making where stereotypers can pick up the new work in a matter of days. Similarly, letterpress machine minders who are experienced on rotary machines have quickly learned to operate web-offset equipment. The effect of the ease of training is that firms are probably more willing to retain and transfer existing workers than they would be if this was much more costly than recruiting ready-trained workers. However, it is perhaps more important that recruitment to the skilled trades depends on the attitude of the trade unions, and the use of internal labour resources avoids this problem.

Where redundancy was unavoidable, the firms in our cases all paid

319

compensation to those involved, even though the redundancies preceded the Redundancy Payments Act. There was no general agreement on redundancy arrangements in the industry, and it was left to firms to make their own arrangements with or without consultation with the unions. In one case mentioned above, employees received 4 weeks' basic wages plus 2 weeks' basic wages for each year of service, while in the other case of newspaper closure payments were graduated from 1½ to 3 weeks' basic wages depending on service. In one case where a series of plant rationalizations was in prospect the firm had drawn up standard formulae for redundancy and resettlement payments, but their apparent absence in most cases was probably attributable largely to the unions' general reluctance to countenance redundancy. The frequent use of very short periods of notice of closure again meant that *ad hoc* redundancy schemes were more likely.

III. THE GENERAL PROBLEMS OF LABOUR-FORCE ADJUSTMENTS

(i) *The Causes of Job Redundancy*

All our cases were concerned with technological change, but there are other sources of job redundancy. One is structural change, either of a general kind where an industry is in decline or where a particular firm is becoming less competitive; a second is a cyclical change in production which may lead a firm to declare workers redundant; and a third is 'policy' change, where withdrawal of Government assistance or orders creates unemployment or where rationalization or take-over creates redundancy which might not otherwise have occurred.[1] In addition to these four inter-related sources of job redundancy—technological, structural, cyclical and policy—there are two dimensions of redundancy, 'unit' closures, where a complete plant or factory goes out of business, and 'sectional' closures, where only a part of the labour force becomes redundant. The point about introducing these distinctions is that different types and magnitudes of change have different effects on job redundancy.

As a general rule, technological changes are likely to take a considerable time to plan and implement and this allows labour force changes to be forecast and planned. Other types of change may be less foreseeable, and structural changes may initially be attributed to temporary and reversible factors so that advance notice of labour-force cut-backs is short. Changes in Government policy may similarly cause unforeseen labour-force problems, while rationalization of a

[1] Policy changes are, in a sense, accelerated structural change.

company or group may be forced as a defensive action which has to be taken more swiftly than would be the case with a technological change. The longer the period of notice of a labour-force cut-back, the easier should it be to use the early adjustment devices—wastage, early retirement, and stoppage of recruitment—to minimize worker redundancy. However, in our cases there was some kind of threshold period before which employees were unlikely to react to news of a closure. Though the closure of the unit was sometimes announced 18 to 24 months in advance, it was only about 6 months before closure that wastage and limitation of recruitment became important. Thus, even in situations where the period of notice is quite short, these early adjustment devices could be significant, but the planning of their use and the organization of an orderly labour-force rundown may require a longer period of warning than can be given in a structural or policy closure, and of course in some cases the notice given is measurable in weeks rather than months.

The size of the closure is important too, since early adjustment devices will be more important in a unit closure than in a sectional closure, in which there is a possibility of internal transfer and redeployment, their success depending on the proportion of total unit employment affected. The labour-force adjustment problems are likely to be least serious of all in technologically induced sectional closures since firms introducing a technological change are more likely to be able to control the speed of introduction of the change and may in any case be expanding output and employment. Technological sectional closures were particularly heavily represented in our case studies, and even where whole units were closed there were frequently other units nearby so that transfer was a possibility. In a unit closure where the employer has no other employment available within daily travelling distance transfer is not possible and the main contribution to reducing worker redundancy must therefore come from the early adjustment devices. For them to be effective and efficient in a structural or policy closure may mean continuing the plant in operation for a period longer than may seem warranted on economic grounds. In a unit closure, once the early adjustment devices have operated, the only alternative to redundancy is resettlement, and we must now consider whether the experience of our cases in this respect accords with other evidence.

(ii) *Resettlement*

In our study, the only really successful case of resettlement was in the chemical industry, where over 100 men agreed to remove to another labour market area, and in other cases offers of resettlement were accepted by few of the workers. Resettlement, therefore, had little effect in reducing the number of worker redundancies since workers were not in general willing to move to more distant plants with job vacancies, and this conclusion is supported by the evidence of other studies.

An early survey of redundancy practices, by the Acton Society Trust, found that a number of employers 'stressed the unwillingness of workers to accept jobs in the same company which involved moving their home',[1] even though a few firms had been successful in persuading workers to move. Wedderburn's investigation of redundancy in two railway workshops showed that only 140 men out of 1,385 were transferred from one workshop, and no one apart from apprentices from the other, which employed about 300.[2] Of those who said they had had an opportunity to apply for transfer, 31 per cent had not done so because it meant moving house, and 32 per cent because it was 'too far to travel'.[3] In a closure of works by the British Aluminium Company, generous financial assistance and the provision of housing did not induce hourly-paid employees to move to another plant: only about 1 per cent finally agreed to move.[4] The Social Survey Report on labour mobility between 1953 and 1963 asked a large sample of workers whether they would be prepared to move with their present employer to another part of the country, and 28 per cent of non-apprentice hourly-paid workers gave an unqualified affirmative.[5] This proportion seems remarkably high in view of the actual evidence on resettlement, but the explanation may simply be that, as several of our companies found, the proportion of employees expressing a genuine interest in moving was very much higher than the proportion which finally did move.

Of course there are exceptions to the generalization that employees are unwilling to move with the employer, and in some instances the numbers moving have been large in absolute terms though not large relative to the numbers leaving for other reasons. Since 1962, the National Coal Board has operated a scheme to provide financial aid

[1] Acton Society Trust, *Redundancy: A Survey of Problems and Practices* (London, 1958), p. 14.
[2] D. Wedderburn, *Redundancy and the Railwaymen* (University of Cambridge, Dept. of Applied Economics, Occasional Paper No. 4, Cambridge University Press, 1965), p. 69.
[3] Ibid., Table 5.8, p. 135.
[4] A. Fox, *The Milton Plan* (Institute of Personnel Management, London, 1965).
[5] Amelia I. Harris, *Labour Mobility in Great Britain 1953–63* (Government Social Survey, London, 1966), p. 32.

and housing for miners who transfer from declining to expanding pits or coalfields.[1] Between April 1962 when this Inter-Divisional Transfer Scheme was introduced and March 1967, 8,801 men had been resettled, and a follow-up study estimated that over 80 per cent adapted well to their new job and area.[2] The number transferred in 1966–7 was 1,711 but no information exists on how much of the £6·5 million spent on social measures was attributable to resettlement. However, to put resettlement in perspective, it is interesting to compare the 1966–7 total of 1,711 with the total number leaving the industry through other reasons. 19,095 men left the Coal Board's employment for medical reasons, retirement or death; 4,587 were dismissed; 1,750 were declared redundant, and 35,255 went as a result of voluntary wastage. The number of resettlements is therefore very small as a proportion of the total manpower loss, and indeed the 1964–5 Report commented that 'if more men availed themselves of the opportunity to move offered by the scheme the importing Divisions could have accepted them' (para. 139).

Though some companies have found hourly-paid workers willing to move, successful resettlement schemes involving reasonably large numbers of workers are much more common for white-collar workers, who showed in the Social Survey report a much greater willingness to consider movement with the employer. This unwillingness of wage earners to accept resettlement is simply one manifestation of the general reluctance to move to another area. This has been demonstrated by many labour mobility studies. Wedderburn found that only 26 per cent of her sample of railwaymen had even considered moving to find a job,[3] while in Kahn's study of redundant car workers half of those who had been unemployed for more than two months had not considered looking for work outside their home area.[4] More than half of the Social Survey sample said that nothing would induce them to move right away from their present area of residence to find work,[5] while confirmatory evidence from many countries suggests that not only are the money costs of movement and the problem of housing important, but also that the 'psychological costs' of leaving

[1] For details see National Coal Board, *Annual Reports*: a summary of the 1965 position can be found in Wedderburn, op. cit., pp. 178–9, while some general information appears in Solomon Barkin (ed.), *Technological Change and Manpower Planning* (OECD, Paris, 1967), pp. 207–14.
[2] This information and that following is taken from the National Coal Board, *Annual Report*, 1966–7, pp. 34–5.
[3] Wedderburn, op. cit., Table 5.11, p. 142.
[4] Hilda Kahn, *Repercussions of Redundancy* (Allen & Unwin, London, 1964).
[5] Harris, op. cit., p. 23.

a home area are very considerable. Consequently, employees will often take a less good job rather than move and will often return to the home area after a period away.[1]

Of course, although resettlement is generally ineffective, it was quite significant in one of our cases, and the Coal Board scheme has managed to resettle an average of 1,500 workers per annum. However, the significant point about the Coal Board scheme is the wide range of benefits offered. These suggest that the total cost could be over £1000 per transfer (more if housing is provided) and against this kind of expenditure the amounts by which even generous companies are prepared to subsidize resettlement look meagre indeed. Few companies could afford or would want to afford this kind of expenditure. In particular cases resettlement may enable the employer to fill chronic vacancies and to offer continued employment to workers who are valuable to the firm for some specific skill they possess and whom it is worth while subsidizing, but the evidence suggests that few companies offer subsidized resettlement to hourly-paid employees, and few employees accept. In the light of the general unwillingness of employees to move to a different labour market area to accept new jobs, it is difficult to see resettlement as a very important part of public or private manpower policy. Indeed, from an economic point of view resettlement would appear to be a particularly inefficient adjustment device. The cost of resettlement is hundreds or even thousands of pounds per head and its success, as measured by the willingness of employees to stay in the new area, seems normally to be limited. It would seem that the money might be better spent on increasing redundancy pay, in improving early retirement terms, or in providing more adequate retraining allowances. However, social considerations are important here both in national and company policy, and in our cases resettlement was viewed not so much as an economic proposition as a social obligation the companies had decided to accept.

(iii) *Transfer and Retraining*

Transfer of employees within the company was remarkably successful in our cases in helping firms adjust their labour forces to changing circumstances, mainly because of the type and structure of our cases. First, the demand for labour was growing in most companies as often happens when a cost-reducing technological change is introduced into

[1] For discussion of these points, see International Labour Office, *Unemployment and Structural Change* (Geneva, 1962), pp. 93–9, 166–7; and L. C. Hunter and G. L. Reid, *Urban Worker Mobility* (OECD, Paris, 1968), pp. 53–7.

firms where output is growing or is expected to grow. In several cases in the steel and chemical industries, companies were able to forecast expansions in production and employment and were prepared to 'carry' potentially redundant workers until they could fill emerging vacancies. Secondly, most of the cases were situated in areas of high employment, and normal wastage during the period of rundown and closure enabled many employees to find alternative employment elsewhere; transfers were also possible to unaffected plants where normal wastage left job vacancies. Thirdly, the characteristic type of change was a sectional closure. Even where units were closed, in a number of cases there were several plants within a limited geographical area and transfer was possible without resettlement or movement of house. In all our three industries there was at least one case of this kind, and it was particularly important in chemicals. However, the ease of transfer depends not only on the existence of job opportunities, but on the occupational pattern of vacancies and of employees who are to be transferred. If the employees are not immediately able to perform the vacant jobs they may have to be retrained before transfer, but retraining was not a significant element in transfer in our cases and we must consider why this was so.

There are various levels of retraining for different types of worker. For maintenance or craft workers, retraining at its most radical would mean the acquisition of a completely new skill, but it is most uncommon for such retraining to occur, since a skilled man has a considerable investment in his craft and the acquisition of a new skill would be costly for the individual and his employer. Also restriction of entry to skilled trades usually prevents the retraining of adult workers. More common is the acquisition of new techniques within the craft, or the ability to work with new materials. This form of modernizing craft training was the most advanced type of craft retraining encountered in our cases except for a few individual cases where electricians and fitters were retrained as instrument mechanics. In fact, most craft workers in our industries could be redeployed to vacant jobs without the need for retraining, since the demand for skilled workers was high in most of our case-study areas and in the printing industry recruitment was restricted by trade union entry regulations. If craft workers who were transferred to a new plant had to acquire some new experience, the most elaborate procedure was a short period of formal training, followed by a period of plant familiarization while the plant was manning up and job familiarization during the running in period: the length and complexity of training depended on the sophistication of the new plant compared

325

with the old. However, it is not often possible to assess the sophistication of retraining, since the broad trade-group names, e.g., fitter, electrician, plumber, are insufficient to show real changes in the work actually done, even though it is obvious that the workers are exercising a different aspect of their craft.

For production workers, there are again different levels of retraining, ranging from small alterations in the existing job duties to a complete change in the work which is done, but again in our industries retraining for production workers was often the relatively straightforward adaptation of existing knowledge, partly because the training was not difficult to acquire and partly because companies tended to select the most suitable workers for particular jobs. In the chemical industry, the best qualified and most suitable workers were promoted or transferred to the new plants, with their jobs being filled by the next most suitable, and so on: thus job familiarization and on-the-job training was generally sufficient, though there were examples of short formal training courses for senior operators on new plants. In the steel industry, promotions and downward transfers took place on a more structured basis, with employee movements confined to relatively narrow job bands, e.g. crane drivers, rolling-mill crews. Within these bands movements took place in strict accordance with seniority. Thus an upward movement was a single step-up to a job which a man had been understudying, while a downward transfer within a job band normally meant a move down the hierarchy to a job which the worker might well have done already on his way up the ladder. In both cases, very limited training was necessary though movement outside the job bands, already barred by institutional restraints, might be very difficult because of the quite specific nature of a workers' experience. Again, however, there are exceptions. In one case where electric arc furnaces were installed, an elaborate theoretical and practical training course was organized for the most senior process workers, but this radical type of technological change and so complex a training procedure is unusual. In the printing industry where the craft tradition is strong and where the conventional job allocations are closely guarded, there were no examples of extensive retraining. As in the other industries, there was no possibility of transfer between craft and non-craft employment. Most non-craft jobs were relatively simple and retraining therefore did not prove difficult. Movement between craft jobs did occur without significant retraining problems. In one case this was because rotary letterpress machine minders were transferred to web-offset machines and the basic element of skill—handling reels of paper being printed at speed—was common to both. In another case, where stereotypers were transferred to making offset plate-making, the new technique was 'deskilling', and retraining was unnecessary.

The evidence of our cases is therefore that technological change creates no problems of retraining and since this conclusion would appear to be at odds with the conventional view of how technological change affects the labour force, we must attempt to reconcile what we have found with what other studies have shown. The key fact here is that our studies dealt with *company* employment policies and the companies' use of their *internal* labour market in a situation where they had spontaneously introduced technological changes over which they had a good deal of control. The companies either had vacancies or created vacancies into which they were able to fit workers for whom retraining proved to be a relatively simple matter because of the redeployment methods used, or because of the nature of the skills required, or because institutional regulations prevented complex retraining (which might not in any case have been worth while). The favourable situation in the internal labour market and the stress on company employment policies tends to understate the problem of retraining. In types of change where the demand for labour in the company is not increasing, particularly structural or 'policy' changes, unit closures or rationalization, retraining may be an essential part of re-employment policy, though it does not then emerge as a *company* adjustment device. For if the company has no vacancies, it may be forced or it may choose to use wastage and redundancy to release on to the external labour market workers whose occupation or job experience may not fit them for immediate re-employment. Then they will either have to move to a different and probably less satisfactory occupation, or they will have to be retrained to fill vacancies in occupations which are in demand. But the problem of retraining in the external labour market no longer focuses on the action of individual companies. The symptoms of the problem appear in changes or imbalances in labour-force structure, and its solution lies in labour market policy to correct these imbalances. Viewed in this way, retraining is obviously much more important than we have so far acknowledged, and the next chapter will discuss it in more detail with reference to changes in labour-force structure and public policy.

(iv) *Redundancy*

In our cases, the extensive use of the early adjustment mechanisms—wastage, early retirement and stoppage of recruitment—and of transfer within the company meant that the number of worker redundancies was minimized, and the effect of these redundancies on individual employees was mitigated by the prevalence of redundancy payment schemes. All companies in our industries had payment schemes which were generous by the standards of the time, and though it is difficult to compare our companies directly with the 'average' of the time, they do appear to have been in advance of normal practice. The little available evidence on the coverage of redundancy schemes comes from the Ministry of Labour survey of 1963 which showed that over 250,000 employees in the Chemical and Allied industry (about 45 per cent of the total) were covered by redundancy policies, and 210,675 were entitled to receive extra notice and severance pay.[1] In Metal Manufacture 124,000 out of about 600,000 employees were covered, while for Paper, Printing and Publishing only about 11,000 employees were covered by some kind of redundancy policy, and most of them did not receive extra notice of a closure. In 1963, then, it appears that the number of workers protected by established redundancy policies was relatively small in the steel and printing industry, much larger for chemicals. All our cases post-dated this survey, but no later information is available which might show in general how widespread redundancy policies were. However, for the steel industry, further information has been provided by Smith, whose international comparison of redundancy practices found that of fourteen British steelworks, only half had permanent redundancy policies; some of these dealt only with the methods of treating redundancy and made no provision for severance pay.[2]

It is particularly interesting that most of our companies in steel and chemicals already had schemes in existence and were not forced to introduce completely new *ad hoc* schemes to cope with a particular situation of redundancy. A common-sense explanation for the incidence of established schemes is that they are more likely to be common in industries where technological change occurs continuously and where the advance of techniques is rapid, since in the absence of a redundancy policy there is likely to be frequent dislocation of production and labour-force problems. Again one might expect to find, as Smith did, some association between the amount of actual redundancy and the existence of established redundancy policies,[3]

[1] *Ministry of Labour Gazette*, February 1963, p. 52.
[2] A. D. Smith, *Redundancy Practices in Four Industries* (OECD, Paris, 1966), Table 2, pp. 70–1.
[3] The association is not very close in Britain if one considers all industries. For example, neither the shipbuilding nor the motor industry were in the lead with redundancy schemes, though both had a good deal of redundancy.

since the need for a scheme either *ad hoc* or established, then becomes recognized. Smith commented:

'to some extent, the rapidity with which treatment has followed the eruption or precipitation of a redundancy problem reflects the frequent use of "ad hoc" redundancy schemes in the United Kingdom' (p. 89).

Of course an *ad hoc* scheme introduced to deal with one situation may be continued in operation and become permanent and established. In the chemical industry, the continuous nature of change is well-recognized and doubtless explains the prevalence of redundancy schemes, but it is impossible to say whether chemicals and steel have suffered a disproportionate amount of redundancy, since no statistical data exist for the periods in which our cases occurred. Since then, the only data come from the operation of the Redundancy Payments Act, but this covers only employees with more than two years' service and so is of very limited use, since many redundancies affect short-service workers.

The distinction between established and *ad hoc* schemes is not clear-cut, for in both the steel and chemical industries where established schemes existed the companies freely amended their schemes to fit them to the needs of a particular situation. This affected both the policies on the dismissal of workers and the payments which they received. The voluntary redundancy schemes in chemicals are one example of *ad hoc* introduction of particular practices to facilitate labour-force adjustment, and there were also several types of special payment, for example, retention payments or specially generous redundancy pay for employees who remained in the employer's service until closure (chemical and steel industries); revision of redundancy pay to take account of transitionally low earnings (chemical industry) or high earnings (steel industry); and supplementary unemployment benefit for workers who remained unemployed (steel industry). There was also in our cases an awareness of the problems of phasing redundancies and of ways of releasing workers gradually to suit the local labour market situation. These techniques included closure of non-essential parts of the production unit, cut-downs on maintenance, contracting-out of certain ancillary functions, and extensive use of temporary labour.

Fundamentally, then, redundancy caused fewer problems than might be expected because most firms made extensive use of other adjustment devices, and they had redundancy schemes which were modified to suit the circumstances of the case. But our conclusions on redundancy, like those on retraining, cannot be generalized since again we are dealing only with technological change, and we must conclude by making some allowance for this, and by assessing the overall importance of redundancy in technological change.

It has been estimated that the annual number of redundancies in Britain is between 250,000 and 1 million, out of a total number of job terminations of $10\frac{1}{2}$ to 11 million.[1] This total of redundancies contrasts sharply with the number of redundancies for which compensation under the Redundancy Payments Act was given in 1967, of about 235,000; it seems probable that the majority of redundancies are of workers with less than two years' service as one might expect under conditions where seniority or last-in-first-out is a major criterion of redundancy. But there are two important points to bear in mind: the first, which really needs no elaboration, is that many redundant employees go immediately to new jobs, and the second is that relatively few redundancies are directly caused by technological change, as was noted by an ILO study:

'despite the fact that in processes or plants directly subject to technical change reductions in labour requirements per unit of output often occur, rarely does there appear to be any net reduction of the labour force employed in the enterprise as a whole. Even less frequent are actual dismissals due to technical change.'[2]

This of course is the conclusion reached in our study, where transfer and the early adjustment devices were successful in most cases in minimizing the amount of redundancy, and where firms were able to control the introduction of technological changes and do a good deal of pre-planning.

However, we must again distinguish the immediate effects and the secondary effects of technological change on the labour force. Redundancy creates relatively few problems for firms introducing technological change, because they can use the early adjustment mechanisms, and because they usually have sufficient time to be able to redeploy labour in the *internal* labour market. Again the problems for public policy arise when one considers structural change, rationalization, cyclical changes in demand, or the secondary effects of technological change, since here there may be no new employment opportunities in the firm and events may move too quickly for an

---

[1] Ministry of Labour, *Dismissal Procedures* (HMSO: London, 1967), p. 57.
[2] International Labour Office, *Unemployment and Structural Change* (Geneva, 1962), p. 23.

orderly rundown of the labour force. Also there will be employees who cannot easily find another job in the locality, because of declining demand for their particular skill or experience, and this presupposes the need either for retraining to combat occupational immobility or an active resettlement programme to counteract geographical immobility. In large unit closures, all employees must find work in the external labour market, but even where only a partial labour-force cut-back takes place, the firm may be unable to re-employ workers within its internal labour market. Where circumstances were less favourable than in our cases and where local labour market conditions and company policy were different, redundancies might be declared much more readily.

When redundant employees enter the external labour market there are two problems: they must maintain their standards of living during any spell of unemployment, and they must find re-employment. Public policy has dealt with the former through the Redundancy Payments Act, but less has been done in terms of positive labour market policies to assist re-employment. Chapter 13 will resume discussion of these issues, but from the point of view of company employment policy the Redundancy Payments Act seems to have had an effect on both employers and workers. Employers seem to have been more willing to declare redundancies, and since up to 1968 their financial liability was limited to the 22 to 33 per cent of the payment not rebated, there was an incentive to use voluntary redundancy schemes to cut back the labour force. Employees apparently became much more willing to volunteer or to stay with the company until closing date in order to collect redundancy payment. The result was that in 1966–8 the number of redundancies under the Act was well above the Government's earlier forecasts and the level of payment was higher, which meant that workers being declared redundant were older or had longer service than had been anticipated. There is no evidence as to how far companies were declaring redundancies without attempting to redeploy workers within the company and how far employees were refusing transfer and deliberately choosing redundancy because of the money they would collect. A strict interpretation of the Redundancy Payments Act might disentitle an employee to benefit under these latter circumstances, but if the employer chooses to allow the employee to go, there is little that can be done about it. One of the interesting findings of our cases in both the printing and chemical industries is the degree to which the employee was given the option of re-employment or redundancy with compensation, and while this would minimize the hardship to the employee it may not be a particularly economic way of dealing with a labour-force cut-back. As Smith commented:

'in some situations, certain safeguards may be needed to protect the employer, and indeed society as a whole, from incurring unnecessary redundancy costs.'[1]

Where a worker is uneconomically re-employed, the firm and its efficiency will suffer; where he is allowed to become redundant though he could have been re-employed, the Redundancy Fund and the taxpayer bear the main burden.[2]

The Redundancy Payments Act goes some way to preventing financial hardship for those employees covered, but several problems remain. First, the Act does not cover any employees with less than two years' service, irrespective of their age or total labour-force service. The Act only deals with a minor proportion of those becoming redundant—perhaps about one-third of the total—and if length of service continues to be the main criterion in determining who becomes redundant, the redundancies are likely to continue to affect mainly short-service workers who receive no redundancy pay. Secondly, a fundamental problem of the Act is that it treats payment as compensation for job-loss, and does not relate any part of the payment to ease or difficulty of finding new work. It is assumed that redundancy pay will enable an employee to be more careful or selective in looking for a new job, but no evidence exists as to whether it actually has this effect. Earnings-related unemployment benefit obviously also has this effect, in theory at least, and one can argue that part of the expenditure on redundancy payments would be better spent on re-employment measures. Thirdly, the Redundancy Payments Act has little effect on company redundancy *policy*. Though the Contracts of Employment Act lays down minimum periods of notice for dismissal, there are few regulations on policies which should be followed when redundancy occurs, except that employers who intend to claim rebate under the Redundancy Payments Act are required to give at least 21 days' notice of redundancy involving more than 10 workers. In 1968 the Department of Employment and Productivity issued a booklet *Dealing with Redundancies* which outlined the principles and practices which ought to be followed in a situation of impending redundancy. The question is whether these advisory services are sufficient. It would be most unfortunate if the

[1] Smith, op. cit., p. 93.
[2] Since early 1969 the employer has had to bear 50 per cent of the total redundancy payment.

degree of financial security now conferred on workers led employers to be less concerned about the methods used in avoiding redundancy if possible, or about following enlightened policies in any necessary dismissals.

Our case studies and other evidence show that the adjustment of the labour force to all types of change can be most easily effected if employers are able to give notice of the labour-force consequences, so that early adjustment devices—wastage, suspension of recruitment, early retirement—can be used, and so that transfer, resettlement and the phasing of redundancies can be planned in some detail. If on the other hand companies simply give short notice of closure so that a large number of workers are thrown on to the labour market at the same time, the task of manpower policy in re-employing them—not easy at any time if there is 'structural' redundancy—will be made much more difficult. It is obvious too that redundancy policies and payment schemes are much more easily concluded at the level of the company or plant, especially since the provisions of schemes may require to be altered to fit the circumstances of particular cases. The Royal Commission on Trade Unions listed the concluding of agreements covering the handling of redundancies as one objective of plant negotiations,[1] and the Commission for Industrial Relations may provide pressure towards a nationally approved procedure for dealing with redundancy. One would expect this to be much more effective in improving procedures than present methods which rely largely on publicity.

Our conclusion, then, is that it is the efficient re-employment of redundant workers which should be the aim of private and public policy, and that such re-employment takes time to organize and plan. In our cases, with their stress on technological change, with the favourable internal and external labour market situations, and with the relatively progressive manpower policies, internal re-employment was generally successful for those employees who had not already found other jobs. But in other situations imbalances in labour-force structure may occur, and it is then the task of public policy to carry out the adjustments which lead to re-employment.

[1] *Report of the Royal Commission on Trade Unions and Employers' Associations*, para. 182.

CHAPTER 13

TECHNOLOGICAL CHANGE AND MANPOWER POLICY

1. INTRODUCTION

The overall impression that emerges from the previous chapters, and especially those dealing with our three industries, is that the companies introducing large technological changes have generally been able to cope adequately with the ensuing adjustment problems. Certainly there were some difficulties. Labour-force adjustment plans did not always work as had been expected or hoped, and in some cases, no doubt, managements faced with the same problems for a second time might follow different lines of action. But the general success of policies and procedures adopted might lead one to question the concern shown in some quarters about the employment consequences and adjustment problems of technological change in the present phase of economic and industrial development. Indeed, some of the conclusions of the case studies can be generalized more widely, since they appear to be fairly representative of the broad trends in the economy as a whole. The growth of white-collar employment visible in our industries is a more general phenomenon, and so too is the change in occupational structure consistent with the changes in technology, in particular the growing use of more automatic methods of production incorporating more sophisticated control and information systems. This has meant a shift away from employment of low-skilled manual workers and towards the employment of more highly skilled and qualified labour, yet in our case studies the transition was effected with little disruption and few difficulties of training or redundancy.

We must recognize, however, that such changes in the occupational structure need adequate adjustment procedures, not only at the level of the companies introducing technological change, but also at the level of the labour market as a whole. The concern for public policy must be whether changes in labour demand induced by changing technology are likely to impose a major additional burden on the labour market's adjustment processes, either quantitatively or

334

qualitatively; that is to say, does technological change add unduly to the volume of labour seeking new employment, and does it give rise to a compositional imbalance in demand and supply due, for example, to the systematic displacement from jobs of workers with one kind or level of skill, while opening up job opportunities elsewhere which demand quite different sorts of skill?

On this broader aspect of adjustment to technological change, the evidence of our case studies is deficient. Although the adjustment devices used by companies in the study were largely successful in dealing with the problems of transition, several problems remain. In the first place, successful adjustment policies in respect of the labour force as a whole can nevertheless mean that some work groups and individuals still have adjustment problems. For example, older workers were more badly affected by redundancy, and in many cases the continued employment of some such workers was regarded as a social obligation which other companies might not be willing to accept.

Secondly, the companies studied were mainly large and progressive, and were experiencing a reasonable rate of growth. We did not study an industry undergoing a structural decline in employment or output, and here the adjustment problems of technological change might be much more serious. New employment opportunities may be limited, and may be concentrated in occupations different from those in surplus. Hence there may be considerable redundancy and the need for occupational mobility if prolonged unemployment is to be avoided

A third and related point is that the cases investigated were in companies actually introducing the change or with control over its introduction. These firms were able to plan the adjustment to change over a relatively long period, and to use many internal adjustment devices to mitigate its employment effects. The more serious consequences of technological change may, however, often be in firms which are not themselves undergoing such change, but which suffer the secondary effects of change in other firms, for example by being unable to innovate or otherwise keep up with their more progressive competitors. Their share of the market may decline, with a consequential fall in employment which may be unanticipated. This problem is not one which can be handled, as in most of our cases, by the adjustment mechanisms of the internal labour market. It depends for its solution on the flexibility and efficiency of the labour market as a whole.

These are the aspects of technological change which are most likely to require special attention in the form of a developed manpower policy, designed to improve those parts of the general adjustment process of the labour market which have proved to be least effective in the face of change. A great deal has, of course, already been achieved in the evolution of governmental manpower policy, though much of this is of recent origin. Until the early 1960s, the view of government was predominantly one of leaving individual companies to get on with tasks such as training, the handling of redundancy, and compensatory arrangements while providing a framework in which government did its best to manage the level of employment overall and in which certain services (such as the public employment service, payment of unemployment benefit, etc.) were supplied to assist the working of the labour market and alleviate some of the hardship which followed from imperfections in the system.

The recent changes have not been aimed solely at easing the problem of adjustment to technological change, but many of them have some bearing on this. However, before discussing manpower policy in detail, we should note that in the mid 1960s Government policy became much more interventionist in several respects. For example, in 1966 a deflationary policy had as one of its ostensible aims, the 'shaking out' of labour from areas of overmanning and its redeployment into high-productivity firms and industries, especially in the export sector. Secondly, the introduction of the Selective Employment Tax in 1966 was intended partly to check the rapid rate of growth in service employment relative to manufacturing employment. Thirdly, there has been in the years since 1964 an attempt to apply an effective incomes policy, and an apparent acceptance of a higher level of unemployment than in the 1950s has been partly offset by vigorous efforts to prevent serious regional disparities in unemployment. Thus, although one of the main aims of manpower policy is still to achieve a balance in the labour market— and in particular to avoid continuing shortages of certain types of labour, notably in the skilled and white-collar groups, and potential surpluses in the unskilled groups—there is, as we shall see, a much wider range of measures available to the Government.

The changing labour market environment which these policies represent is relevant to the ease with which adjustments to technological change are likely to be effected. The possibilities of redeployment within firms or among firms, industries and occupations will be governed in part by the general condition of the labour market and by the pressures acting on labour demand in different sectors of the economy. Governmental influence on the overall structure of employment, and on the growth and structure of wages, either through its general policy or through more specific measures, may affect the scope and direction of labour mobility, the employment policies of companies and even the bargaining attitudes of trade unions concerned with the consequences of technological change. Furthermore, if there is a bias in the use and displacement of particular kinds of labour due to technological change, the direction of *other* forms of economic change—including those emanating from Government policy—will be important.

In the present case, other economic forces affecting the structure of industry seem to be moving the demand for labour in the same direction as technological change. In view of this, the possibilities of maladjustment in the wage structure due to incomes policy and income standstills, and the tendency to raise the acceptable level of unemployment, may make it rather more difficult to achieve ready adjustment to the displacement effects of technological change and to redirect both new labour-force entrants and displaced personnel towards the expanding occupational sectors of the economy. In these circumstances, it becomes more than ever important that other aspects of manpower policy should be examined to see how far they are capable of correcting for some of the actual or potential difficulties.

II. POLICY FOR SHORT-RUN ADJUSTMENTS

This section is concerned with the response of manpower policy to the need for workers affected by technological (and other) forms of economic change to adapt to the new requirements of the labour market. It thus deals with those aspects of policy which relate to employment and income security and other provisions which are likely to affect the worker's financial position during the transition to new employment, the facilities for adapting the skills of displaced workers to match with those currently in demand, and the improvement of information services in the labour market.

(i) *Income and Employment Security*

We have already suggested in Chapter 12 that one of the tests of successful adjustment to technological change is the minimization of actual worker redundancy. Ultimately, therefore, what would seem to matter most is the improvement and dissemination of techniques for dealing with technological and other forms of change, to provide a greater degree of employment security. Ideally, this would mean that firms should anticipate changes in technology in organization

337

and in the availability of labour, and plan changes in the level and structure of their employment with the minimum of disruption either to production or to the existing labour force. Although government, mainly through the Department of Employment and Productivity, has begun to improve knowledge of forecasting techniques at company level, it has had little apparent success in advising companies about methods and practices of increasing employment security, and legislation has had to be introduced.

The Contracts of Employment Act of 1963 ensured that reasonable periods of notice of termination of employment were given to employees with more than six months service, the period of notice increasing with length of service. This provided assurance to a major part of the labour force that they would have a fair warning of impending redundancy, giving them a chance to start looking for new work, but of course it did little in itself to oblige employers to improve practices relating to employment security. On a rather different point, government has also indirectly contributed to a reduction in casual employment (for example in the docks), where insecurity of income and employment is perhaps greatest. But the major problem would still seem to be the insecurity of workers who regard themselves as a regular part of the work force of a company and who have no guarantee that their employers will choose or be able to adopt 'enlightened' techniques of effecting changes in employment by means of forecasting and planning.

The main emphasis of policy has in fact been on income security, which is admittedly a rather easier area in which to legislate, though as suggested earlier, this kind of emphasis could have the effect of making companies less concerned about methods of avoiding redundancy.[1] The two main innovations in recent years have been the Redundancy Payments Act and the provision of an earnings-related unemployment benefit. The first of these provides severance payments for workers made redundant through a variety of causes, with eligibility depending on a minimum of two years' service with the company, and payments being a function of age, service and recent wage levels. These are much the same criteria as adopted by companies operating compensation schemes prior to the Act, but the compensation is more generous and only half of the cost of payments falls upon the company, the remainder coming from the national Redundancy Fund. Under the Act, payments are in the form of lump sums and are made whether or not the worker obtains

[1] Conversely, of course, it would be argued that more emphasis on employment security could have added to the overmanning problem.

a new job immediately. Its rationale, therefore, is very much that of providing compensation for a break in employment outside the worker's control. If the worker is unable to find a new job, he is eligible for unemployment benefit or, if he enters a Government Training Centre, he receives a grant during training. The second innovation is that those with recent earnings above a specified level receive unemployment benefits (and training allowances) geared to their earnings, thereby cushioning them against the serious fall in income that would result from a drop to the standard rate of unemployment benefit.[1]

Both innovations are relevant to the adjustment to technological change. Redundancy payments, under the Act, are guaranteed to workers with at least two years' service, where previously they were non-standardized and far from universal. This seems likely to have some beneficial effect on trade union and worker attitudes to the implementation of technological change, provided that clear efforts have been made to minimize worker redundancy or to allow the brunt of redundancy to fall on those for whom the problems of finding new work are likely to be least severe. The earnings-related unemployment benefit, too, has advantages to the higher paid workers who suffer redundancy and cannot immediately obtain new employment. Together with redundancy compensation, this should make it possible for displaced workers to 'shop around' for new work, rather than take the first job offered, and the prospects of satisfactory employment and efficient use of such labour may be improved. There is, however, no guarantee that the money will be used in this way by the unemployed worker.

These innovations, then, have standardized and formalized practices operated on a piecemeal and erratic basis prior to the new legislation. Their contributions seem on the whole to be beneficial to the adjustment to technological change, but they leave one or two problems which are worthy of consideration. Successful redeployment within a factory may involve both the transfer of labour to jobs carrying higher earnings and equivalent or higher status, and the transfer of workers to jobs which are lower paid and lower in status. In the latter case, workers will mainly have accepted the prospect of lower pay as part of the price for continued employment. For some the fall in earnings may be slight, and it may be only a temporary lapse while a person becomes established on a new

[1] For the first six months of unemployment, supplementary payments amount to about one-third of the claimant's average weekly earnings between £9 and £30, subject to a maximum *total* benefit of 85 per cent of his earnings.

promotion ladder, but as we have seen in some of the industry studies, workers may be prepared to accept a permanent reduction in earning power towards the end of their working lives, acknowledging their lack of suitability for absorbing new training or their willingness to undertake less arduous and responsible work. Nevertheless, there will be some for whom the drop in income will be substantial and long-lasting, and it can be argued that they are as deserving of compensation as others who are actually paid off—especially those who can be fairly sure of immediate re-employment. A form of compensation was paid to workers in the Samuel Fox case discussed in Chapter 7 above, where the company acknowledged responsibility for reducing the earnings expectations of some of its workpeople through technological change. Whether this is something that is directly amenable or appropriate to public policy is debatable. It may be more a matter for managerial discretion or perhaps more aptly for negotiation at company level within the context of agreements relating to redundancy procedures, especially since general criteria are not easily laid down and since individual workers' prospects and circumstances are relevant to any settlement.

Altogether, it is difficult to see how public policy can reasonably do much more to give security of income to those affected by technological change, either through lump sum compensation or regular income maintenance. The main problem is to ensure that needless disruption to employment security is avoided, and that efforts to improve security provisions do not inhibit labour mobility that is essential to the economy. From this point of view, apart from the need for many companies to have access to information on techniques of forecasting, planning and redeployment of manpower, the principal area for development may be that of company personnel policies. This is one of the issues taken up in Section IV below.

(ii) *Retraining and Job Placement*
One of the most obvious effects of technological change is that it tends to create a need for workers to alter and improve their skills. As our industry studies have shown, the responsibility for retraining is frequently accepted by the companies responsible for the change, partly because it is often easier to adapt the skills of existing employees, partly perhaps because of difficulties in securing new labour already equipped with the necessary skills, and partly in some cases because the firm has a social conscience which prevents it from simply declaring workers redundant and leaving them to find new work

340

outside the firm. Even so, there will be many cases where it is not possible to find all existing employees new work, or where the firm (especially the smaller company) has no facilities for retraining or scope for redeployment, or where workers themselves are unwilling to accept alternative employment within the company and wish to improve their skills and seek other types of work. There will also be cases where firms are obliged to close down altogether, leaving no possibility at all for redeployment and retraining in the firm's own internal labour market. In such cases, the provision of alternative facilities for retraining which cannot be provided privately must be accepted as a public responsibility.

The main effort in this field has been a major expansion in the number of Government Training Centres (GTCs), the most important function of which has been the training of adult workers (though the bulk of adult training is still done by private industry).

In 1962 there were only 13 GTCs in existence. By 1968 the total had risen to 38, and the target for 1970 is 55, with an output of 23,000 trained workers per year, most of whom will have undertaken six-month training courses in particular trades, either to replace a skill that is now obsolete, or to acquire a higher level of skill than before.[1] This is undoubtedly a useful contribution to the retraining problem, though there are qualifications. First, by no means the whole of the projected 23,000 trainees annually will be unemployed adults undergoing retraining. Secondly, are the facilities offered in GTCs 'adequate'? A proper measure of what is 'adequate' is almost impossible to determine. A great deal depends on the extent to which workers being made redundant are actually in need of retraining, on the characteristics of the redundant workers and their suitability for training under the criteria laid down. Thirdly, it has to be remembered that in most industries retraining is carried out by industry itself. Despite all this, the trends in occupational structure indicate that the need for retraining facilities in GTCs may be greater than even the projected capacity. Even though there is little evidence of queuing for entry to the Centres, much more could be done. Indeed, the lack of pressure on the GTCs may simply reflect the fact that they are doing a reasonable job within the limits of their present structure. However, these limits are rather narrow, and the apparent adequacy of GTCs is conditioned by the requirements which are laid down for entry to GTCs, the general reluctance of many unemployed workers

[1] In practice, it seems that a high proportion of workers training in GTCs are people who have voluntarily left jobs in order to acquire a skill: the proportion of such workers may be as high as 40 per cent of the total.

to consider entering a Centre for training, and the composition of the courses provided.

Each of these factors is relevant to the future impact of technological change on the employment structure. First, many of those likely to be made redundant by technological change will be in the lower skilled sectors of the labour force and, perhaps in part because the Redundancy Payments Act is making employers more willing to lay off older workers, the average age of these displaced workers will tend to be higher. The problems of providing training for older, low-skilled workers are well known, but if they are to comprise a larger share of the redundant in future, as seems probable, the GTCs may have to face up to the difficulties, using specially developed methods of adult training. Secondly, the attitude of the workers to retraining and indeed to all types of further training as a recurrent part of the working career may have to undergo radical change. There is no *necessary* presumption here that the progress of technology is so speeding up that enforced changes in job will become more frequent, though this may happen. However, it does seem that technological change, by expanding the relatively more skilled types of work at the expense of the less skilled, is enlarging the portion of the working population in which a continuous process of relearning and updating of skills is characteristic.[1] Thirdly, the courses provided in GTCs encompass a relatively narrow range of skills. This takes us on to the major topic raised earlier, the State's ability to introduce flexibility into the whole conception of job structure and training requirements.

As the Royal Commission Report reaffirmed,[2] there are frequent obstacles to the employment as skilled workers of persons trained at GTCs, and inevitably where such difficulties exist there will be a tendency both for Centres to select courses from which trainees can go on to use the skills acquired, and for potential trainees to be unwilling to accept training for jobs which might subsequently be barred to them, or in which they might, as dilutees, be virtually 'second class workers'. In this area, there is perhaps relatively little that can be done in the short term beyond government pressures upon those trade unions whose attitudes to job-definition, demarcation and entry through apprenticeship alone are antagonistic to the

[1] It is true of course that the semi-skilled worker may learn to perform operations on a number of different machines in the course of his working life, but this is much more a process of job-familiarization than a real case of changing or improving a skill. See also below, Section IV.

[2] Cf. Royal Commission on Trade Unions and Employers' Associations, *Report* (Cmnd. 3623, HMSO: London, 1968), p. 89.

development of an enlarged and more varied retraining programme. Given the kinds of technological change that are now occurring, we must in the longer term envisage the demise of the craft system of training which can no longer be regarded as the most efficient method of training skilled labour. Yet the craft system, depending on property-rights to particular types of work, is one which is well defended against erosion and the decline may be slower than desired.

There is of course another aspect of placement, relating partly to those who have undergone training but partly also to those who have been displaced by technological change but for one reason or another are not eligible for or interested in retraining. Many of those who become redundant will find jobs on their own initiative, but many will also seek assistance through the local employment services of the Department of Employment and Productivity. The activities of these services are too well known to require spelling out, and there has not been any marked change in the role they play. However, in a number of cases of redundancy on a substantial scale, local employment service officials have established interviewing facilities within the works or in some other way obtained direct access to the redundant workers before the redundancy occurred rather than waiting until the workers became unemployed and registered with the local exchange. The ability of these services to help in placement of course depends on the state of the local labour market, the demand for the skills and experience the redundant workers have to offer, and the ability of the company to give advance warning of impending redundancy. A company which intends introducing a substantial technological change should have ample opportunity to warn the local office, since this is not the kind of change that comes about suddenly and unexpectedly. Even where large numbers of workers are thrown out of work with relatively short notice, as when accelerated structural change affects a company or factory, a co-ordinated exercise by the company and the DEP may often have remarkably successful results.

In summary, then, we can conclude that public policy towards the problems of short-run adjustment to technological and other forms of economic change has altered considerably in the last few years. However, attention cannot be confined only to these short-run problems. Is the labour market system as a whole capable of meeting the demands made on it by changes in the economic structure of the economy, many of them stemming directly from technological advances which affect the industrial and occupational composition of the demand for labour? The next section takes up this question.

III. POLICY FOR LONG-RUN ADJUSTMENTS

In this longer-term perspective of manpower policy, there seem to be three main items for consideration. First, there must be some attempt to forecast the likely long-term development of manpower requirements if policy is to be able to concentrate on areas of potential difficulty. Secondly, one of the most important means of adjusting to these long-run requirements is the existence of an educational and training system geared to providing supplies of labour capable both quantitatively and qualitatively of satisfying the trends in demand. And thirdly, particularly in a full employment society where there is a persistent shortage of certain types of labour, concern has to be shown for the efficiency with which labour is actually used. This brings into the scope of long-term policy the ability of the industrial relations system to secure an efficient use of labour resources and its capacity for facilitating the transfer of these resources when the occasion arises.

(i) *Manpower Forecasting and Planning*

Apart from some interesting efforts at manpower planning under wartime conditions and immediately post-war, British governments had not seriously involved themselves in this area until the 1960s. Much of the effort, however, has been of a primarily short-term nature, as for example in the National plan of 1965 and in the work of various Committees concerned with forecasting demands and supplies of certain key groups of workers.[1] Our main interest is in attempts to predict the longer term developments in labour demand, and here the work of the Manpower Research Unit of the Department of Employment and Productivity is most relevant. The Unit was set up in 1963, one of its main tasks being to study the probable trends in manpower utilization and their implications for labour market policy. In practice, it has concentrated on manpower analysis and forecasts on an industrial basis, though *within* the industries studied the occupational aspects have received attention. This is not, however, the same as a more global analysis of employment by occupation or occupational group, since the industry-by-industry approach does not show up the possible surpluses or deficits of

[1] Notably the Committee for Manpower Resources for Science and Technology, set the task of overseeing the utilization of qualified manpower (engineers, technologists, scientists and technical supporting staff). Various *ad hoc* bodies have reviewed the situation in other occupational sectors such as medicine and dentistry.

particular types of labour in the economy as a whole—which is what really matters when consideration is being given to the content of training programmes designed to prevent shortfalls in the nation's supplies of essential manpower. In other words, it will usually be rather easier to arrange a redistribution of an adequate supply of a particular type of labour than to procure an increase in the supply, the problem becoming progressively greater the more specialized and highly qualified the labour in question. Obviously, a great deal more needs to be done in this general field of longer-term forecasting and planning on an occupational basis, though such efforts may have to await the development of more adequate occupational statistics on the existing labour force, and greater research into the work content of occupations themselves and the inter-relationship between occupations in terms of work-content and potential for mobility.

There is, however, another aspect of public policy in relation to forecasting. Just as the economy as a whole needs to have knowledge of changing trends in labour utilization, so also does the firm need such information as a basis for its training and recruitment policies. Up to a point this is a matter for the firm itself, and may be relatively short term by nature, probably no more than five years and often less. However, although as we have seen in the industry chapters the larger firms seem generally able to make forecasts that are adequate to meet the needs of new investment programmes, it is by no means certain that medium and small firms are equally well equipped for this task. Here again, public policy may have a longer-term role to fulfil in developing awareness and know-how of forecasting techniques and applications—though the relevant techniques may themselves relate only to shorter-run problems at plant level.

The Manpower Research Unit, in its Report on the metal industries found that:

'less than one in four of all firms approached said they made some sort of forecast for more than two years ahead, and only one in forty said that this covered all categories of worker. In all, about half of the firms said they did some kind of forecasting, but in the majority of cases this was merely a one year look ahead.'[1]

How far the average firm really needs to engage in forecasts for *all* its labour is questionable, since there may be fairly large sectors of its employment which are either subject to little change or can be

[1] *The Metal Industries* (Manpower Papers No. 2), Ministry of Labour, HMSO: London, 1965. The Report also observes that the sample inquiry was biased towards the larger firms.

expanded or decreased without any major planning effort to avoid shortage or prevent hardship on the part of the workers. However, in many companies, even the small ones, there arc likely to be some types of labour worthy of a systematic study from the viewpoint of future demand and supply. Adequate publicity about general trends in utilization of different types of labour may help, but even then there may be a lack of knowledge on forecasting methods and national trends can only give a broad picture which is not typical of individual areas and largely irrelevant to the average firm. There may then be a case both for attempting forecasts of occupational trends at local level and for informing companies how forecasting of their own labour situations on a selective basis might be helpful in preparing for the future.[1] The first of these tasks could be most suitably undertaken by local offices of the Department of Employment and Productivity, though it would necessitate some addition to their staff resources. The second, being a task that requires both a knowledge of techniques and access to companies and their labour force problems, could be undertaken by employers' associations or by the Department of Employment and Productivity.

Forecasting is only a means to an end. It produces information giving advance warning of possible difficulties and often enables the rough dimensions of future imbalances to be determined. Forecasts can be revised and rolled on as new information comes to light. But in the last resort it *is* only information, and where problems are identified action still has to be taken. Apart from retraining, which has already been discussed as a method of improving the short run situation, attention must be paid to the adequacy of the training system as a whole. In this area, as we now see, government policies have also been evolving in recent years.

(ii) *Training and Education*

The adequacy of the facilities for training people to meet the need of industry for a wide spectrum of skills cannot be considered in isolation from the nature of the broader base of the educational system as a whole. The main concern here, however, is solely with

[1] A recent publication by DEP has helped to fill this gap: cf. Company Manpower Planning (Manpower Papers No. 1), HMSO: London, 1968. The point is strongly made that manpower forecasting exercises: 'should concentrate as far as possible on critical skills or occupational groups—i.e. those where shortages are most likely to upset the company's programme, where recruitment is most difficult, where the period of training is longest, or where labour costs have been rising steeply' (p. 12). This type of approach by DEP may, however, need to be supplemented by more direct assistance.

346

TECHNOLOGICAL CHANGE AND MANPOWER POLICY

the stresses imposed on the systems of training and education by the changing character of technology, and with the response of public policy to these stresses.

The current phase of technological change appears to be putting a number of strains on a system of preparation for employment that has not operated altogether satisfactorily in the past. In the first place, technological change seems to be moving the demand for labour in a direction which requires a higher average level of training, and demands a change in the balance between 'on-the-job' and 'off-the-job' training in favour of the latter.

Although generalization is liable to result in over-simplification, it is possible to identify three main types of job preparation. There is the training given to semi-skilled and unskilled labour, usually no more than a brief period of job familiarization but sometimes extending over several months. This training will be almost entirely job-related, with a minimum theoretical content, relatively independent of general educational performance and conducted mainly on the job. Secondly, there is the training given to 'skilled' labour, the most readily conceived example being the time-served craftsman.[1] Training in a skill will normally be a mixture of on-the-job training and class-room instruction either within the plant, in technical colleges under day release or like schemes, or through evening classes. School or other qualifications are only infrequently a formal condition for entry to skilled training. Thirdly, there is training for what may be called 'high level manpower', including managerial and executive personnel, the professions, and a variety of scientific and technical workers. In these cases, school qualifications and commonly degrees or diplomas from a course of higher education are the basis of more specific training for the kind of job in question.

This third category depends most heavily on the educational system for the job preparation of its new entrants. While those with the necessary educational qualifications will need some period of job familiarization, the facilities required for an expansion of supply in this category are likely to be concentrated in the fields of secondary and higher education, rather than provided by industry. The import-ance of this lies in the kind of changes taking place in the occupational structure. The main employment expansion in the past has been in the manual worker category, for whom the main locus of training has been the workplace. But now increasingly greater reliance is being

1 Many workers normally classified as skilled, of course, have not served an apprenticeship but progress to this category by accumulating experience and qualifications in a series of preparatory stages, both on the job and off the job.

placed on the educational system to produce adequate supplies of people with qualifications that can be directly employed or readily adapted by users of high-level manpower, while increasing use has been made of technical colleges to provide parts of the training for craft and other skilled labour. In short, the occupational changes that have been occurring and seem likely to continue involve fairly extensive changes in the orientation of both the educational and training systems, with a switch in emphasis from on the job to off the job preparation. At the same time, there has probably been a complementary change within the training system itself, moving towards increased use of off the job instruction for skilled manual workers.

The second kind of strain on the training system arising from current trends is the growing probability that job content and job attachment will change markedly over a person's working life. In these circumstances it is important that training should be reasonably flexible and not over-specialized towards a static and narrowly defined conception of the job. For many types of worker, the career is a series of job progressions involving at least minor changes in occupational affiliation and requiring recurring extensions of earlier training. However, for manual workers, especially craft workers, it is commonly expected that the acquired skills will continue to be applicable without serious modification throughout the normal working life: the concept of training in such cases is basically of a 'once-for-all' experience. In view of the changes in the occupational structure, however, it seems likely that this conception will diminish in importance. Much more relevant would seem to be the development of training systems (for manual and other workers) providing a basic core of job-related knowledge on to which can be grafted courses leading to specialization, updating of knowledge and enlargement of the area of job opportunity.

How far, then, has public policy recognized the problems and what steps have been taken to improve the situation? The retraining provisions have been discussed in the last section, though it has to be recognized that these measures do not necessarily go very far to changing the attitudes to retraining as a normal part of work experience. Also important here are the attempts by government to introduce more flexibility into the industrial relations aspects of training, job definition, and working practices, to which we come shortly. The main advance, however, has been the introduction of the 1964 Industrial Training Act, by which government has established an organization for regulating the quantity and quality of training provision on an industrial basis, with the main attention falling initially on training for new labour-force entrants entering skilled employment.

The Industrial Training Boards present the opportunity for much greater guidance of industrial training than has hitherto existed, and allow at least minimum standards to be established. This is an important advance in a system which has mainly lacked any real uniformity of training principles or standards. For the present purpose, however, the importance of minimum standards is that employers hiring workers trained under the system will have a more accurate knowledge of the kind of service they are purchasing. Additionally, the new system, if properly linked to forecasts from bodies such as the Manpower Research Unit of requirements for new trainees, should enable the provision of training places by industry to be more closely geared to demand. And finally, a surer knowledge of the constituent elements in a given training such as an apprenticeship, and of the standard of achievement that can be expected from those who have undergone particular forms of training, should make it possible for additional or alternative skills to be moulded on to the existing base. There are, therefore, evident advantages in terms of adaptability and mobility to be derived from the new approach in the longer run, but there are also problems.

The main problems of manpower planning tend to lie in the occupational rather than the industrial affiliations of workpeople. But the Training Boards are based on industries or industry groups, leaving open the possibility that different standards for a given occupation might be set by different industries and thereby leading to a lack of that transferability of skill between industries that is so necessary a part of the adjustment procedure consequent upon technological change.[1] One answer may well be in the designation of certain basic sections of training, to which could be added supplementary sections dealing with the particular requirements of the various user industries: any need for transfer between industries at a later stage could then be met by means of shorter training courses specific to the requirements of the receiving industry. The basis for such a scheme has been laid in the proposal of the engineering industry Board that the first-year training of apprentices should be on a 'module' system, in which a variety of courses can be combined in different ways to produce differentiated skills, with the possibility

[1] This problem has been recognized by the Central Training Council, which has outlined a procedure for co-ordinating the work of Boards as it relates to occupations common to a number of industries. Cf. CTC Memorandum No. 7: *Training Standards for Occupations Common to a Number of Industries*, 1968.

being left open that additional modules can be tacked on subsequently.[1]

In conclusion, it is evident that major changes in the system of training have been taking place in such a way as to reflect the changing needs of the economy for trained labour. It will, however, be many years before the total labour force will comprise a majority of workers trained under the new system which in any case has still some way to develop before it can be regarded as being properly geared to the changing requirements for trained labour and the improved standards required of training. Even if these problems are resolved, however, there remains the important question of the effective area of employment for those who have been trained or wish to embark on a course of training or retraining. There is no point in devising improved training schemes along the lines discussed above if there are barriers restricting the mobility of labour. The problems of training and retraining are by no means entirely composed of deficiencies in the number and structure of training places themselves, but may lie equally in obsolete attitudes to training or in an out-of-date conception of job-definition, while institutionally based schemes for the protection of income and employment may likewise frustrate attempts to redesign systems of training.

(iii) *Industrial Relations*

In an important sense, the aspects of public policy discussed above, from manpower planning to security provisions, have a bearing on the industrial relations environment within which technological changes have to be introduced. Under different environmental conditions, the problems of implementing and of adjusting to technological change might be very different. However, the interest in this section is mainly in the implications for technological change of interventions by the State in the conduct of industrial relations. Of these, the two most prominent at the present time are the pursuit of an effective incomes policy in which the need to relate wage and productivity improvements is a matter of emphasis, and the recommendations of the Royal Commission on Trade Unions and Employers' Associations—particularly that the current dominance of industry-wide collective bargaining should give way to a plant or company based level of negotiation on all but the most general conditions of employment. As we shall see, these are not entirely

[1] The Central Training Council have also set up committees, on occupational lines, to advise on the training of groups found in all industries: e.g. commercial and clerical workers, training officers, etc.

independent issues, and while they raise many questions which cannot be discussed here there are several points of direct concern to the present study. We begin by considering the implications of an incomes policy which favours and indeed actively encourages productivity bargaining.

The productivity bargain has been described:

'as an agreement in which advantages of one kind or another, such as higher wages or increased leisure, are given to workers in return for agreement on their part to accept changes in working practice or in methods or in organization of work which will lead to more efficient working.'[1]

Such agreements are most frequently developed at plant level or at least at the level of an organization capable of conducting comprehensive and detailed negotiations on terms and conditions of employment. The decision of management to enter into productivity bargaining may derive from any of a wide range of factors, but one of these may well be dissatisfaction over current standards of manning, with overmanning and consequent underemployment of labour leading to loss of productive potential and unnecessarily high unit costs. As we have seen at various points in preceding chapters, the issue of manning standards is one that is frequently raised when technological change is introduced, and a large part of the overmanning that is generally acknowledged to exist in British industry can undoubtedly be traced back to failure on the part of management to achieve the lowest standards of manning consistent with efficient and safe operation of newly introduced plant. Even with the more sophisticated techniques of devising manning standards that are now available, teething troubles and other factors may lead to target standards being exceeded, and manning becomes established at a higher level by custom and practice. From that point of view, productivity bargaining can be seen as one means of retrieving a situation of overmanning resulting from technological change.

More importantly, however, technological change by breaking into established practices opens up a new bargaining situation, and in an economy in which incomes policy and the relationships between pay and productivity improvements are given prominence, the scope for managements to achieve levels of manning closer to the optimum is likely to be increased. They may affect the level, and also in some cases the structure of manning. Technological changes frequently

1 *Productivity Bargaining*, Royal Commission on Trade Unions and Employers' Associations, Research Paper No. 4 (HMSO: London, 1967), p. 2.

351

demand that traditional demarcations and the associated training should be cut across, but it has proved impracticable on many occasions to introduce the necessary flexibility of work and workers. Productivity bargaining, and the stimulus given to it by incomes policy, have made changes of this kind relatively easier, though there are areas where customary rights are still strongly defended.

To sum up this part of the argument: productivity bargaining has been encouraged by the course taken by incomes policy, and the publicity given to productivity agreements has caused a shake-up in traditional conceptions of property rights in work, in notions of acceptable and unacceptable degrees of job flexibility and hence in attitudes to technological change. The two main effects have been first, that many companies have managed to derive the benefit of savings in manpower and unit costs created in previous years by technological change; and secondly, that as further technological changes are introduced, the more favourable attitudes and the removal of protective practices make it more possible for companies to achieve standards of manning and modes of operation which are closer to the optimum and which could not have been achieved under the conditions formerly existing.

In the Royal Commission Report there are three main points which relate to the procedures for the planning of technological change and the ensuing adjustment problems. The first is that a factory agreement, as envisaged by the Commission, should include the constitution for a factory negotiating committee which would be empowered 'to deal with work practices within the factory, so that it can negotiate changes in working practices to suit existing or proposed production methods'.[1] The intention here comes very close to formalizing what has actually happened in the negotiation of productivity agreements, particularly those of the package deal variety, for under the traditional industry-wide form of bargaining the constitutional mechanism for this kind of negotiation does not normally exist. In practice, as was evident from the industry studies, many of the changes accompanying innovation have been introduced very largely as a matter of management prerogative, and although certain matters, notably wages on the new plant, come up for negotiation, this is frequently carried out according to procedures determined on an industry-wide basis, and involving union representatives from outside the factory, rather than by a negotiating body within the factory, familiar with the specific problems of the workplace. The exceptions to this exist mainly in companies which are not party to industry-

[1] Royal Commission *Report*, op. cit., p. 40.

wide agreements and which have their own internal negotiating bodies: thus for example in the chemical industry, both Shell Chemicals and ICI were already equipped with negotiating committees capable of dealing directly with the effects of technological innovation. Some further exceptions may exist where the rates determined at industry level are accepted as being minimum rates and where, accordingly, the machinery for supplementary bargaining by shop stewards at workplace level is well established—this would be true, for example, in some printing establishments. Nevertheless, as a general proposition, the development of negotiating machinery within the plant, not bound by general agreements on wages and conditions at industry level, could help to ease the introduction of technological change, in that it would be able to take into account the specific conditions of the factory itself and the detailed effects in the factory of individual innovations.

It would be possible to include in the factory agreement procedures for dealing with redundancy and redeployment, and this is the second point arising from the Commission's recommendations. Some companies do of course have established policies on redundancy procedure, but many of these policies are formulated by management on behalf of the company, and unions do not have the right to negotiate in these matters, though they may be consulted.[1] The Royal Commission was evidently in favour of agreements covering the handling of redundancy, advocating this as one of the objectives that boards of companies should have in mind when reviewing industrial relations.[2] It is difficult to determine how much this would improve the process of adjustment to technological change within the plant. Chapter 12 showed that many companies even in the early 1960s did have prepared procedures for dealing with redundancy, and in our case studies these seem to have operated well. On the other hand, there was in our cases a degree of *ad hoc* treatment, in that the approach to the problem varied from plant to plant within a company, and the details were at the discretion of individual plant managements provided the proposals fell into line with the broad philosophy of the company personnel policy. It is not at all certain that trade unions or workers in the company were aware of the procedures that would be followed in the event of technological change, and some of these

---

[1] Likewise, before the Redundancy Payments Act the payment of redundancy or owner compensation was at the discretion of management, as was the formula for calculating it. Consultation with unions sometimes took place, but no negotiation.

[2] Royal Commission *Report*, p. 45.

procedures, notably those affecting the selection of men to be made redundant and the selection for retraining, were matters of very immediate concern to the workpeople. Also, these cases were within the larger companies, most of which did have a broad view of personnel policy and all of which possessed management teams with personnel specialists and considerable backroom assistance. How far the smaller companies, and even the larger ones with a less well developed personnel policy, would be capable of evolving procedures leading to such successful outcomes is highly questionable. In any event, a good deal can be said in favour of a procedure known in advance, and agreed by unions, for probably the greatest problems in the handling of redundancy and related matters arise out of lack of knowledge, rumour, and what may be unfounded fears about the procedures to be followed.[1]

The third point to be made is in part a development of the two points already discussed. One of the consequences of a shift to company level collective bargaining, as foreseen by the Commission, is the removal of the protection afforded by the present system to boards of directors:

'At present boards can leave industry-wide agreements to their employers' associations, and the task of dealing with workers within those agreements to their subordinate managers. Removing this protection will direct the attention of companies to the need to develop their own personnel policies.'[2]

The precise content of a personnel policy will vary from case to case, but it will normally pay considerable attention to matters of recruitment, training, promotion and welfare arrangements including the range of fringe benefits. So far as technological change is concerned, the points that would seem to matter most are those relating to the opportunities provided to existing personnel to train for and be promoted into what are often the high pay and prestige jobs opened up by innovation. Additionally, it will be important that the personnel policy should indicate the preparedness of the company to plan its labour requirements in the light of anticipated changes in demand and changes in technology so as to minimize the incidence of worker

[1] Some substantiation of this view comes from the steel industry where the principle of promotion by seniority is well established among process workers. Though the system may have disadvantages elsewhere, it does provide a known system of procedure in dealing with redundancy and the related redeployment of labour. This is not a recommendation for the seniority rule itself, but is a sign in favour of an established and well-understood method of procedure.
[2] Royal Commission *Report*, p. 41.

redundancy. Where policies exist, they will not be static, nor will they be inflexible, for inevitably a concern for the problems and conditions of individual workplaces must be shown. Once again, these are responsibilities accepted and acknowledged by many of the largest companies, which give a clear line of policy to be administered in detail by factory managements, but undoubtedly there are a great many more that have no clear conception of what personnel policy they are pursuing. As the Commission correctly observed, a move to plant or company bargaining will not by itself achieve the development of more sophisticated personnel policies, but if it does contribute to such a development some of the problems of change may be diminished or more easily handled.

IV. CONCLUDING REMARKS

The discussion of this chapter has illustrated the growth of State interest and intervention in labour market policy, and although the implications of technological change are only a part of the overall picture, it is clear that within the last decade the range of State influence in this area has grown substantially. The evolution of labour market policy has to be viewed as a whole, and of course the more recent developments are additions to a well-established framework of employment services and governmental encouragement to a voluntary system of industrial relations and collective bargaining.

Even in the limited area of adjustment to technological change, as we have seen, the extension of public policy has had important effects in such varied activities as manpower forecasting and planning, training and retraining, and provision against insecurity in various forms. Elsewhere, as in the renewed emphasis on pay and productivity relationships, government persuasion may have influenced the direction of collective bargaining and, perhaps less directly, the atmosphere of bargaining and the attitudes to the consequences of technical innovations. More recently still, there are the changes being set in motion by the Report of the Royal Commission, and as changes in the system of industrial relations are pushed forward, the approach to the problems presented by technological change may become more sophisticated and of better all-round quality.

On the whole, then, the recent expansion of public policy in this field has gone a great deal of the way towards eliminating or minimizing those deficiencies in the system that are amenable to governmental influence. Some of the existing institutions established by government will of course mature and expand, as in the field of

M*   355

industrial training, but these will tend to be autonomous developments rather than the consequences of government directives. If there is a gap which seems to require filling it is in the provision of information and technical advice to companies, particularly those of medium or small size in which the resources available for specialized personnel functions are frequently meagre. The Commission for Industrial Relations may help to fill this gap, but on the manpower side it is doubtful if enough is yet being done. It is again debatable whether this requires action by the Department of Employment and Productivity, or whether assistance might be more readily provided by employers' associations acting as quasi-consultancy agencies. It may well be that there is a function for both, with the Department developing techniques for appraising manpower situations at national and local level, and others helping individual firms to interpret and apply the resultant information to their own circumstances.

Given the time lag that exists between the industry studies and the date of publication, we cannot expect that this volume will present ready and immediate answers to specific problems of technological change. This in any case was never part of our purpose, and in the event public policy and developments resulting from the enterprise of companies and unions have altered the picture since the basic studies were undertaken. The nature of the problems associated with technological change, though it may have changed in detail and inevitably will continue to do so, has not however changed in broad principle. We believe that this kind of approach can help to identify more closely the problem areas of technological change and adjustments to it and that, although industry differences emerge for obvious reasons, the contrasts that are thrown up, as well as the difficulties faced in common by various industries. are useful guides to a better understanding of the issues.